STACKING THE DECK

STACKING THE DECK

Winning the Personal Finance Game

KEVIN IKENO

STACKING THE DECK
WINNING THE PERSONAL FINANCE GAME

Copyright © 2016 Kevin Ikeno.

All rights reserved. No part of this book may be used or reproduced by any means, graphic, electronic, or mechanical, including photocopying, recording, taping or by any information storage retrieval system without the written permission of the author except in the case of brief quotations embodied in critical articles and reviews.

The ideas and strategies in this book do not replace individual financial planning consultation. Prior to implementing any of the tactics detailed herein, you should review your situation with a financial advisor. The author and publisher explicitly disclaim any responsibility for any liability, loss or risk, personal or otherwise, that is incurred as a result, directly or indirectly, of the use and application of any of the contents of this book.

iUniverse books may be ordered through booksellers or by contacting:

iUniverse
1663 Liberty Drive
Bloomington, IN 47403
www.iuniverse.com
1-800-Authors (1-800-288-4677)

Because of the dynamic nature of the Internet, any web addresses or links contained in this book may have changed since publication and may no longer be valid. The views expressed in this work are solely those of the author and do not necessarily reflect the views of the publisher, and the publisher hereby disclaims any responsibility for them.

Any people depicted in stock imagery provided by Thinkstock are models, and such images are being used for illustrative purposes only. Certain stock imagery © Thinkstock.

ISBN: 978-1-5320-0577-0 (sc)
ISBN: 978-1-5320-0579-4 (hc)
ISBN: 978-1-5320-0578-7 (e)

Print information available on the last page.

iUniverse rev. date: 10/10/2016

♠ CONTENTS

Acknowledgments vii
Introduction ix

1. What's Your Plan? 1
2. Personal Financial Statements 11
3. Understanding Credit 37
4. Credit Vehicles 49
5. Savings and Debt Management 59
6. Investing 69
7. RRSPs and Retirement 87
8. Tax-Free Savings and RESPs 105
9. Estate Planning 115
10. Protection 123
11. Taxes 129
12. Owning versus Renting 137
13. A Final Word 147

Appendix 1: Know Your Financial Self 155
Appendix 2: Recommended Resources 183
Index 187

♥ ACKNOWLEDGMENTS

Words alone will never be able to fully express my appreciation to my children and their mother.

To my muse, Frances: Things didn't work out between us, but you stood by me through one heck of a roller coaster, and you gave me five amazing kids and sacrificed your career to raise and home-school them. Even after a breakup, you helped me with the creation of this book. I am forever indebted to you and in awe of your generosity. Thank you for being my inspiration when I needed it and my kick in the butt when I deserved it.

And to my kids: You are my motivation.

Raphael, as you are an adult now, I hope these pages help guide you past some of the pitfalls I fell into. Sebastian and Saffyre, if it hadn't been for you guys arguing over Crazy Eights so many years ago, the seed of the idea of this book would probably never have been planted. Dagan and Caoimhìn, fill your childhood with smiles and laughter and enjoy every experience that life offers you.

I would be remiss if I didn't say a special word of thanks to the leaders and mentors I have had over the years I spent in the financial-services industry—most notably, but in no specific order, Glenn Bullock, Mark Kenny, Gerry Doyle and Mike Smith.

To all of my clients over the years, especially to those of you who kept telling me that I should put my ideas, concepts and strategies into a book to share with everyone—in particular, to Marcel Alama, who just kept asking "Is it done yet?"—all of you are the people who kept encouraging me and helping me believe and continue on, even when things got rough.

And finally, to everyone who has bought this book, thank you for helping turn a dream into reality.

INTRODUCTION

An investment in knowledge pays the best interest.
—Benjamin Franklin
US Statesman and Philosopher

Last week I was enjoying some quiet time with the family, and two of my kids were playing a game of cards. The majority of time, they play very well together, and this was one of those times—up until one decided the other wasn't playing by the rules. This initiated a conversation the likes of which many of us have heard or participated in:

"You cheated!"

"No I didn't!"

"Yes you did!"

And you know where that goes.

That evening, I was putting together some material for an upcoming workshop, and I was experiencing a block, uncertain of the right chronology. I was reflecting on the situation between the kids when I realized that the two topics were very closely related.

As an adult playing cards, if you think someone is cheating, it's likely because you feel they are stacking the deck in their favour. In life, many of us may feel the deck is stacked against us because the cost of living goes up at a rate greater than or equal to our income. This includes property taxes, income taxes and fuel costs (even though government inflation calculations seem to omit these). So how can we ever get ahead?

We want to be debt-free. We want to have not only the necessities of life but some of the comforts and maybe a few luxuries as well. But how is this possible with governments, banks, corporations and the other major players in the market holding all the good cards? The banks make billions of dollars in profit, and we want to pay fewer banking fees. The government takes half our income in taxes, and we just want to pay our bills.

It is possible for the big players to keep doing what they're doing and for you to be financially successful. I am going to share a few ideas with you that I have learned over the last decade or so in the financial-services industry that it seems a large percentage of the population is generally unaware of. I believe that knowledge is power, and I am going to help you embark on an adventure that will show you how to stack the deck in your favour. But first, to know how to change the odds, you must understand what game you are playing. Let's start by reviewing the basic principles of the game.

The name of this game is Financial Success. That means different things to different people, but the basic principles are the same:

1. Spend less than you earn.
2. Have a plan.
3. Have a backup plan.
4. Monitor and track your progress.
5. Update your plan at least once or twice a year.
6. Make proactive positive changes as required.

I once heard someone say, "Money isn't everything, but you can't say that without it," and I feel this is a very accurate statement. When you are trying to enjoy life and raise your standard of living, it is impossible to do so without money. That is an undeniable truth.

Imagine your lifestyle (or standard of living) as a snowball you are, throughout life, pushing up a hill. During childhood, you are dependent on family to push that snowball for you. As you grow up and move out on your own, you perhaps rely on the charity of friends and family to a certain degree to help you keep the snowball moving forward, this is the first stage called dependence.

Think about your first apartment or even your first home. In the beginning, maybe you had Grandma's 30-year-old couch; it was dark green and really didn't match other stuff in your place, but that didn't matter because it was comfortable, it was free and it pulled out into a bed for when your buddy couldn't quite make it home after a poker game. During that time, perhaps you were a "starving student," and every so often a family member would come to visit with a bag of groceries—or you spent your weekends at Mom and Dad's house doing laundry and eating your only meals that came from the traditional four food groups as opposed to the postsecondary food groups: bottled, bagged, canned or frozen. This is how many of us spend those years in higher education.

Then school is over, and, other than student loans, you probably don't have a lot of debt, so your monthly debt payments are reasonably affordable, as are other lifestyle expenses. You have graduated from postsecondary and into the next stage of life: independence.

Independence is the stage in which you are able to float in the ocean of life on your own without the big rubber ring of external support around your financial waist to keep your head above water. Progress is slow and may, at times, seem non-existent; but income and expenses are relatively equal. You are able to support yourself without assistance.

Each of these stages all of us experience at different points in our lives, depending on the decisions we make and how well we establish a foundation and plan for the unexpected. We work hard building our careers, and enjoying our personal lives. Then we meet that special someone, buy a house and start a family.

Now you are starting to acquire assets—things you own that have a resale value, because Grandma's couch isn't going to be worth much more than memories. As important as those memories are, they don't help increase your standard of living. At this point, you are starting to enjoy a quality lifestyle.

Over the years, you finance new cars, the mortgage gets paid down more and more and you start to enjoy some of life's comforts. These may include a trip for the family occasionally, a vacation here or there, a newer vehicle even though the old one is not that bad and extra gifts "just because" for those special people in your life.

So you continue to work hard, save, pay off and eliminate debt, perhaps purchase a vacation property and amass other luxuries—things that you want but don't necessarily need. This is the final stage. It is the point at which your debts are virtually gone (except those that provide tax benefits; I will discuss this more later). Each month, the bulk of your income is available for you to do with as you please. Many people, for a variety of reasons, never make it to the comforts stage, let alone the luxuries stage.

By virtue of the fact that you're reading this, you already have one of the most important components in financial success: desire. If you don't want it, aspire for it and do everything in your ability to acquire it, simply put, you won't get it. In anything you do, your level of success is in direct correlation to your level of desire and motivation to succeed.

Many individuals find their desire and motivation challenged when life throws a curveball and they are not prepared. In order to reduce the impact of these curveballs, you must have a plan. We cannot predict the future, but we can prepare for it.

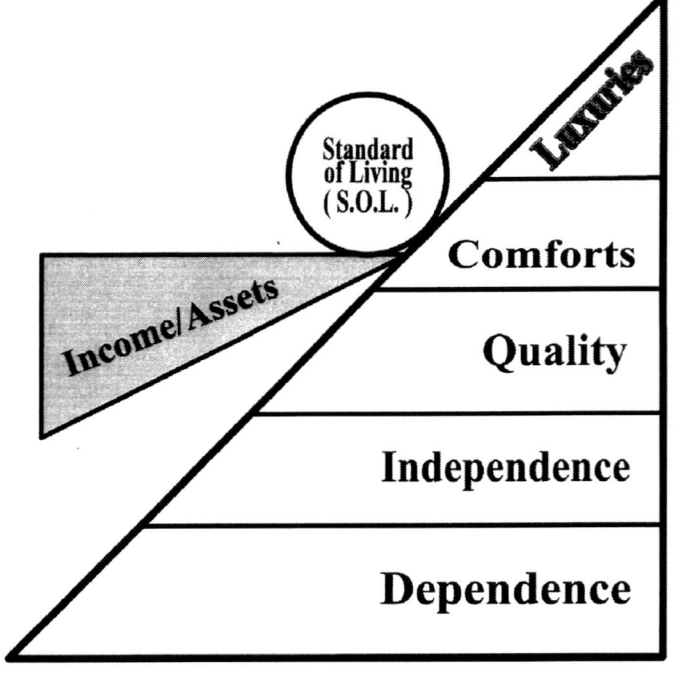

Now this snowball we call our standard of living, which we have been pushing up the hill of life—what holds this standard of living in place? We work 40 to 60 hours per week trying to build a standard of living that will provide our 2.5 kids and our chosen life partner with (what we hope will be) not only what they need but what they want as well. That standard of living is held in place and indeed is pushed forward by our income and assets.

What happens when this wedge of income and assets that we have lodged in place gets hit by a curveball in the game of life—health issues or changes in the economy? Suddenly it's like our snowball has hit a patch of slippery ice, and it starts rolling back down the hill because we have significantly reduced income and assets.

Proper planning for these what-ifs can help secure your snowball in its place on the hill so that it doesn't slide very far, thus making your move up the hill easier the higher you get. Over the years, I have seen many situations where people did not have a plan, and when they started to slide down the hill, they experienced significant financial stress. Sometimes families were torn apart. This book will help you avoid that by showing you where you are right now on the hill and teaching you how to build a strong wedge to improve your odds of being financially successful.

I believe that everyone will find tremendous value in reading what I have learned over the last 15 or more years in finance, especially those people who are

- new to finance/just starting out in life with their first job,
- new to Canada/considering moving to Canada and looking to understand the Canadian system,
- close to retirement and finally getting around to evaluating what their lifestyle will look like when they do retire and
- among the many Canadians who have been motivated to get a better understanding of their finances by the last few years of international economic turmoil.

By following my advice in this book, you will make a great investment in your future.

1 What's Your Plan?

Are these things really better than the things I already have? Or am I just trained to be dissatisfied with what I have now?
—*Chuck Palahniuk*
Author

Canadians today are becoming more conscious than ever of their personal finances. We are coming from a past where there was a lack of understanding, a lack of education and a lack of resources on personal finance. The school system didn't teach us about the subject. Our parents didn't teach us about it—whether it was because they didn't think it was important, they were afraid of the topic or they were ashamed that their own finances weren't in order. Perhaps our parents just thought that if they bought us all kinds of stuff growing up, it would create the illusion that everything was okay, even though they were just going further into debt.

We are in an age now where that has changed, or is substantially in the process of changing. Not only are there more sources of information via newspapers, books, television and the Internet, but also there are more companies offering financial advice and guidance. The demographics of the financial industry and consumerism in general have changed and continue to change radically.

As Canadians, we are constantly trying to find the balance between our work, personal, professional and financial lives. With so much of our time being dedicated to what some refer to as "our corporate masters" and raising our families, it is challenging to find time to enjoy those things which bring us individual happiness—whether that means getting out into nature, spending time with our social circle or simply reading a good book. Oftentimes short-term and long-term financial planning are put on the "someday when I have time" list, and given thought only to the extent of purchasing a home, renewing a mortgage and attending a preretirement seminar five years before the "big day."

If financial success is important to you, take it seriously. Make a plan. Write it down. Review it at least once or twice a year.

Why do I need a plan?

Recent Stats Canada information indicates that one out of three Canadians will suffer a disability of greater than 90 days before the age of 65. Imagine that for three months, you receive only 60% of your current income. Many people have disability protection through their employer, which averages around 60%, and that is why I am using this figure for the example. Now your $2,000 biweekly paycheque is $1,200.

It seems to me that Canadians have perfected the art of living paycheque to paycheque. That means they have expenses close to, or more than, their paycheque. What happens in situations like this? How quickly do we fall behind? How long does it take to catch up on that shortfall?

Consider the potential risk to your assets, some of which must be sold to cover your regular expenses. On top of that, the possible increase in costs associated with medicine or treatment during the disability is phenomenal. This may also cause long-term damage to your credit history. The ripple effect is quite large because on some forms of credit products, particularly loans and lines of credit, the interest rate you receive is based on your credit history. The better your credit, the lower the interest rate the bank charges. Conversely, the lower your credit rating, the higher the interest rate.

Although you may recover from your disability and go back to work, you will now have liquidated assets, thereby reducing your net worth and pushing your goals years further into the future—or possibly making them completely unattainable. Your credit is not in perfect condition, and you pay more money in interest, reducing the amount of money you have to put toward your goals. You have to set them further in the future or take more risk to earn a higher return.

This situation sounds somewhat bleak—dismal even. This is only one example of where things could go without planning and thinking ahead.

Success in anything doesn't just happen; it requires focus, training, learning and a desired end goal. A financial plan is your road map of where you are and where you are going. It details what you want to do, like saving to buy a house, paying off a debt, paying for future education and many other things.

The majority of decisions we make, whether as simple as buying groceries or as complex as buying a new home, have a heavy influence on how successful we will be at achieving our financial dreams. So we must take these things into consideration by incorporating them into our plan. Throughout the coming chapters, I have included tools, resources and strategies that you can use to start a plan on your own, but I also recommend talking to a financial advisor.

Is it difficult to create a plan?

No, it's not. We now have convenient access to online resources, applications available on mobile phones and all the various sources of media at our fingertips 24-7. However, what works for some people will not work for others. A twentysomething postsecondary student can and does make significantly different decisions than a 55-year-old father of three. Many of these resources do not take these things into consideration, and realistically they cannot.

Your financial plan must be as unique as you are. That's why the information in this book is so valuable. You have to understand your finances to be able to create your plan.

Once I create a plan, am I done?

As life changes, so must your plan. Couples are having fewer children and at a later stage in life than in the past. This means that we become accustomed to certain luxuries that are attainable to a DINK (double income, no kids) family. When we start having children, unless contingencies and planning have been put in place to prepare for this life-changing event, we may find that we are borrowing more, saving less or making uncomfortable sacrifices that cause relationship stress and unease in our personal lives. We work more for faster promotions—or perhaps just to stay out of the house. Communication breaks down, and the decision-makers are making individual decisions, not joint. Two separate paths are created, which typically means both people are not getting what they want out of the finances or the relationship.

Sinking further into debt, we source out more borrowing to get back on track, which is usually only a

Band-Aid on a bullet wound. By making changes to our plan as life changes, we can keep the snowball in place on the hill. What we really need is a financial education.

One of the most common statements I hear from people when sharing my concepts and strategies is, "I wish they had taught me this in school." The second most common statement is "This takes work." Yes, it does. If you were hoping for a secret to achieving financial stability without a little elbow grease … sorry, this isn't it. Personally, I haven't found any method that doesn't require some effort, at least not for the average person.

However, before you run back to the bookstore to return this book, let me just say that a financial plan is really not as tricky as many people believe it to be. Like a well-oiled machine, once you have it up and running, it's relatively self-sufficient. You just have to check up on it every now and then to ensure that it's still working the way you want it to. Once or twice a year (the mandate in the financial-services industry is annually), spend a couple of hours reviewing the following:

- Where was I at this time last year?
- Where did I want to be at this point this year?
- Am I there?
- If not, why?
- Where do I want to be at this point next year?

If there is a major life event, like a birth or a death in the family, or if you plan to make a major purchase like a vehicle or another property, you should also review your plan. When you are completing this annual or semiannual review, maybe even consult a professional financial advisor to help.

I feel I should reiterate that: *to help*. Your advisor should be there to guide and teach you, not tell you what to do. Advisors should provide you with options and recommendations and explain the pros, cons, costs and benefits in order to help you make informed decisions about your finances. If you are relying on others to tell you what to do in this respect, you will never understand your finances, and therefore you will find it more challenging to achieve certain levels of wealth.

To begin your adventure, the first thing you need to do is decide what your goals are. For example:

- I am going to save money to buy a house.
- I am going to pay off my debts.
- I will pay my bills on time.
- I will save money in case of an emergency.

Then, you will need to make your goals SMART: specific, measureable, achievable, realistic and time-bound.

For example, rather than saying, "I am going to save money to buy a house," say, "I am going to save $15,000 over the next three years by putting $400 per month into an investment plan." Rather than making a goal of "saving for emergencies," create a plan that clearly identifies what that looks like. A good emergency/backup plan is to have three months of lifestyle expenses saved. If in order to live your current lifestyle you need $2,500 per month in your pocket, then you would calculate $2,500 × 3 = $7,500 and make a SMART goal of: "I will save $100 every two weeks for the next three years until I have $7,500."

Once you have made your goals SMART and written them down, go to your financial institution and set up whatever you need to reach them. Put a reminder somewhere to make sure to check on your plan in one month or three months. Especially when you are first starting out, checking up on your plan more often is important to remind you of what you are doing and why.

Many financial institutions give you the ability to give an account a nickname. When you have this plan set up, give account nicknames like Emergency Fund or My New House. That way, every time you log on to your online banking, you are reminded of why that money is there.

The really important part of this is, don't wait. Do it now. Do it today.

2 Personal Financial Statements

If you aim at nothing, you will hit it every time.

—*Zig Ziglar*

Author

Every successful business has financial statements. Not only is it a legal requirement for tax-reporting purposes, it is also necessary for accountability to shareholders. Financial statements provide a snapshot of where the business is at and allow people to measure the value and stability of the company, among other things.

One company I worked for in the financial-services industry had a slogan: "What gets measured gets attended to. What gets attended to gets done." The company you work for measures your productivity, your results, your hours worked, your time off and every other aspect of your work life. Why? Because it wants to ensure that your efforts are contributing to a successful business so it can continue to be in business.

We see this and deal with it every day of our professional lives, and yet so many of us neglect it or entirely ignore it in our personal lives. The reasons for that are beyond the scope and intent of this book. What I will do is show you how to understand the components of your personal financial life so that you can—either on your own or with the help of some professionals—move your snowball up the hill as far as you desire and safely secure it in place. Even if you do choose to build a team of professionals like accountants, financial advisors and lawyers to help you with this, you still need to understand what they are doing and why. This will ensure that you make better use of your interactions with these professionals by understanding the big picture.

There are two primary financial statements when it comes to personal finance: *cash flow* and *net worth*. These tools are multifaceted. Not only do they help you measure progress month after month, year after year, but they also are what banks use to approve loans, mortgages and all other forms of credit.

The Cash Flow Statement

On the simplest level, a cash flow statement identifies how much money there is, where it comes from and where it goes. Further simplified:

$$\text{Income} - \text{Expenses} = \text{Cash Flow}$$

Cash flow can be positive or negative. Here is a sample of a cash flow statement that you can use to better understand your cash flow situation.

Income	Amount
Salary/wage	
Pension	+
CPP	+
OAS	+
GIS	+
Insurance	+
Investments	+
Total	=

Expenses	Amount
Saving and investing	
Housing	+
Debt payment	+
Groceries	+
Child care	+
Transportation	+
Health care	+
Clothing and grooming	+
Education	+
Leisure	+
Miscellaneous	+
Total	=

Total income	
Total expenses	−
Total Cash Flow	=

This will help you (and the banks) assess your ability to save, invest or take on additional obligations. You should always be tracking your cash flow.

The goal of financial planning is to generate positive cash flow, so that you have money left at the end of the month rather than month left at the end of the money. If you run out of money before the end of the month, you must do one of two things: increase your income or decrease your expenses. Income comes from salary/wages; investments; government payouts like child tax benefits and universal child care; business income for those who are self-employed or other sources like drawing on a loan, taking out a line of credit or selling an asset. Though these other sources of funds can help short-term cash flow issues, if you are dependent on these sources of income, it will almost inevitably lead to significant financial problems. That is why it is important to simultaneously review the Cash Flow Statement and the Net Worth Statement.

The Net Worth Statement

A Net Worth Statement essentially shows how much money you'd have if you sold everything you own and paid off everything you owe. Assets are the things you own that have resale value. Liabilities are what you owe, such as a mortgage, loan or credit-card balance. The equation is:

$$\text{Assets} - \text{Liabilities} = \text{Net Worth}$$

As with cash flow, net worth can be either positive or negative. Here is a template of a Net Worth Statement that you can use to assess your own net worth.

Assets: What You Own	Value
Property	
RRSPs	+
TFSAs	+
Non-Registered investments	+
RESPs	+
Bank accounts	+
Personal effects	+
Other	+
Total	=

Liability: What You Owe	Value
Mortgage	
Loans	+
Line of credit	+
Credit cards	+
Other	+
Total	=

Total Assets	
Total Liabilities	−
Total Net Worth	=

Cash Flow and Net Worth: Working for You

The goal is to ensure that your cash flow and net worth are continually increasing. The sooner you notice a decline in either one, the sooner you can attend to fixing it and the less likely it is to become an issue that is not manageable.

Expenses can be tracked in a variety of ways, but in most financial readings, you will find them broken out into 11 main categories:

1. Savings and investments
2. Housing: rent/mortgage, heat, hydro, property tax
3. Debt payments
4. Groceries
5. Child care
6. Transportation: public transit (including bus, subway and taxi) or personal vehicle (including loan/lease payment, licensing, insurance, maintenance/repair, fuel, parking)
7. Health care: vitamins, supplements, medicine
8. Clothing and grooming: hair care, makeup, soap, shampoo, toothpaste, cologne
9. Education
10. Leisure: dinner out, movie rentals, vacation
11. Miscellaneous

Technically, you can increase or decrease the number of categories as suits your specific situation as long as the expense is not in more than one category. Many people will, for example, lump personal grooming with groceries, as many of these items are purchased at stores that sell both groceries and personal-care products. As long as you know where you are tracking it and one expense is not in two categories, you will do fine.

The more closely you monitor exact figures, the more accurately you will be able to forecast how to get ahead and where the "leaks" are. Leaks typically are the Tim Hortons coffee on the way to work or the little splurge while you are at the mall with the kids. Suddenly you're wondering, "Where'd it all go?" I am not saying don't do those things; I am just saying keep track of it and keep it reasonable. Ensure the leaks don't break the dam.

Money is a commodity. It helps us acquire things we want and need. It is not meant to be hoarded. Life is meant to be enjoyed and experienced, and money helps us do that. If we save all our money, we aren't experiencing life today, and we might not be here tomorrow.

At the same time, I would suggest that the chief cause of unhappiness and failure is sacrificing what we want most for what we want at the moment. It's about balance. In my opinion, we should look at this as lifestyle planning rather than financial planning. Then maybe Canadians would be more committed to planning to create a better future lifestyle while still enjoying today's lifestyle.

This is why we need to be aware of our cash flow and net worth. We also need to know our short-term and long-term needs and goals so that we don't have to sacrifice, we just have to plan. This plan will likely have to include the evil "B" word: *budget*.

What budgeting is really about

Many people have told me over the years that they don't like to budget. I believe this is due to a misinterpretation or miscommunication regarding the definition of the term. These people have told me, "I don't need to budget. There is no reason to penny-pinch. I make a good income." But that's not what budgeting is about.

I don't care how rich you are. Governments of the world work with a budget, and they not only have the ability to increase their income at will by increasing taxes but they also print their own money. You and I cannot print our own cash … well, I guess we could, but considering the consequences, it would not be a recommended strategy to manage our cash flow.

No one has a good excuse to avoid budgeting. So what is a budget? Nothing more than a projected Cash Flow Statement. We look at previous income and expenses to estimate what future months will look like. With this, we create a savings or debt-reduction plan based on current

Cash Flow Statements. When we do this, we discover opportunities to increase our cash flow and ultimately our net worth.

Cash flow is king. When you have a positive cash flow, you have an opportunity to create your own future—and these are the first steps in the journey to get you there. Here is a sample budget you could use to forecast your future months of spending:

Income Source	Amount
Salary	
Commission/bonus	+
Self-employment income	+
Alimony/child support	+
Dividends/capital gains/interest	+
Government payments	+
OAS/GIS	+
CPP/QPP	+
RRIF/LIF	+
Other	+
Total	=

Expenses	Amount
Saving and investing	
TFSA	+
RRSP	+
Non-Registered	+
Housing	
Mortgage/rent	+
Condo fees	+
Hydro	+
Heat	+
Water	+
Gas	+
Insurance	+
Telephone	+
Internet	+
Home security	+
Other housing costs	+

Debt Payments	
Loans	+
Lines of credit	+
Credit cards	+
Other debt payments	+
Groceries	+
Child care	+
Transportation	+
Fuel	+
Car insurance	+
Repairs	+
License	+
Car wash/detailing	+
Parking	+
Public transportation	+
Other transportation expenses	+
Health care	
Insurance	+
Prescriptions	+
Dental	+
Chiro/physio/massage	+
Vitamins and supplements	+
Veterinary	+
Other health care expenses	+
Clothing and grooming	
Clothing	+
Salon/barber	+

Dry cleaning	+
House cleaning	+
Pet expenses	+
Soap and shampoo	+
Gym membership	+
Sports equipment/fees	+
Other clothing and grooming expenses	+
Education	
Tuition	+
Books	+
Pens/paper/binders	+
Other education expenses	+
Leisure	
Cable/satellite	+
Movies/concerts/shows	+
Alcohol/tobacco	+
Magazines/newspapers	+
Books/music	+
Lottery	+
Vacation	+
Miscellaneous	+
Total	=

Total Income	
Total Expenses	−
Surplus/Deficit	=

There are a few ways to improve your cash flow now that you have started to monitor what's coming in and going out every month. Primarily, your choices are to increase income or reduce expenses. You can increase your income by doing the following:

1. Working more hours at your current employer (assuming you get paid for overtime)
2. Asking your boss for a raise
3. Changing employers/careers to something with a higher salary
4. Getting a second job
5. Expanding your business if you are self-employed
6. Drawing a higher income on your investments

For many of us, it makes more sense, is faster and makes a bigger difference in our cash flow if we reduce our expenses instead. In order to decide how to reduce expenses, have a look at your budget, your cash flow and your net worth to determine where you can spend less. You may also want to consult an advisor. Here is an example of how I helped one of my clients.

A few years ago, I met Donna and John. They had come to me quite upset about an NSF on their bank account. This is a fee the bank charges if a cheque or payment tries to go through your account and there are non-sufficient funds. I could understand why they were upset. No one wants to pay $45 to the bank and $45 to the person they wrote the cheque to because of insufficient funds. Donna and John told me this had happened to them about six months prior with a different bank that they dealt with, and they had promptly set up overdraft protection to solve their problem.

So I asked, "How's that working for you?"

Of course, they said excellent. Now, once or twice every month, when they had a cheque or automatic payment go through the bank before the funds were available, the cheque didn't bounce, and they didn't get charged $90 in fees—only a $5 administration fee and less than $2 in interest for the three or four days until payroll was deposited.

"So that sounds like a very viable solution, doesn't it?" I asked.

"Absolutely!" they agreed, and they wondered how quickly they could get this set up at their new bank.

I told them it would take about 20 minutes—if it was the right solution.

They both became perplexed. "What do you mean *if* it's the right solution?" asked Donna. "What other option is there?"

That's when I asked them an important question: *Why?* Why didn't the payment clear? Why wasn't there enough money in the account?

Probing further, I asked if this was an unexpected payment. Were there unusual expenses that month? Did they regularly use their overdraft at the other bank?

"Yes," John told me. "We use the overdraft often. It has really saved us on many occasions, not only the expense of the NSF but the embarrassment as well."

I still had no answer to my real question: Why wasn't the money there in the first place? Especially since they told me it was a car loan payment they'd come up short on. I asked a few more questions and discovered that Donna was a government employee and John was a teacher. Their joint household income was over $150,000 a year. I then asked them possibly the hardest question they had ever been asked: "Where does it all go?"

The room was silent as they stared at each other.

I listened as they slowly began to talk about the fact that they had a good income and it just seemed like each year their income went up but they were constantly just able to make it through the month with little or nothing left. I reassured them that in my world, among the clients I saw, that was the norm. It didn't matter if they made

$30,000 or $300,000 a year. The majority of Canadians have mastered the art of living paycheque to paycheque.

I spent another 30 to 45 minutes helping them list income and expenses and create a Cash Flow Statement and a Net Worth Statement. With their mortgage payment, two car loans and various credit cards, they were paying out over 70% of their after-tax income, which left very little to maintain the lifestyle of an active couple in their mid-forties with two teenage daughters.

I showed them how refinancing the equity in their home would cut their expenses in half, alleviating the need for overdraft protection because now they had cash flow. They briefly argued that this would eat up the equity they had worked so hard to build, and I responded by reminding them that they owed the money regardless. Under a mortgage, they'd have one payment that was easier to manage and track, the rate would be lower, they'd save on interest expense, they'd reduce the overall cost of borrowing and they'd gain the cash flow they needed to provide peace of mind, reduce stress and save while paying down the mortgage more quickly. Structured their way, over many credit vehicles, the debt was going to take 20 years to pay off anyway.

This is just one example. There are a variety of ways this concept can be implemented.

The Goal Triangle

Now that we understand the concept of cash flow management and we've created more cash flow, what do we do with what we have created? We set a goal. In financial planning, there is common reference to a goal triangle. In the middle of the triangle is your goal.

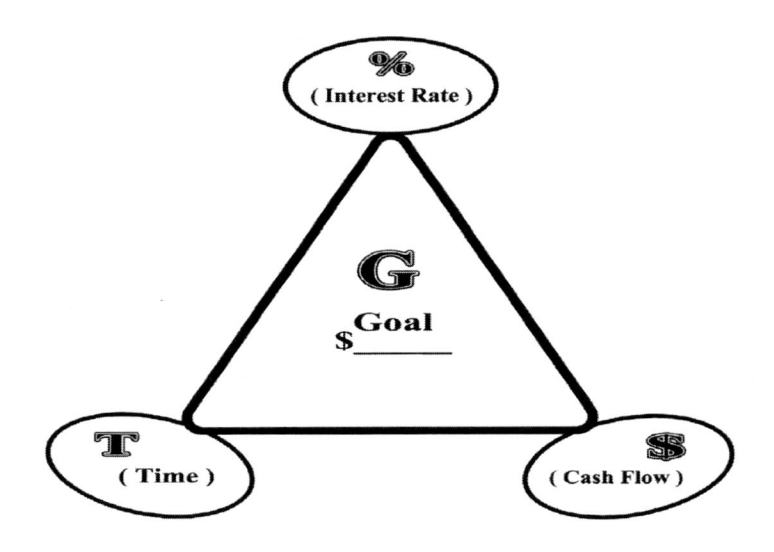

```
          %
     ( Interest Rate )
           /\
          /  \
         /    \
        /  G   \
       / Goal   \
      / $_____   \
     /_____\
   T                $$
( Time )        ( Cash Flow )
```

Virtually everything you do has a dollar value associated with it—whether it's paying off a loan or a mortgage, saving for a vacation or buying a car. The goal in the middle is the actual dollar value that you are working toward saving for or paying off. How quickly you achieve that goal is influenced by three things:

1. The amount of money you are willing or able to commit to that goal; this comes from your cash flow
2. The interest rate, whether return on investment or cost of borrowing
3. Time

Given any two variables, the third can be calculated. If you don't like how the equation works out, change the variables accordingly.

Realistically, the amount of money we have to contribute is often a fixed amount, as our income is our income. One pitfall I have seen with many people is being overly aggressive with their goal and dedicating too much of their cash flow to the goal. This can lead to a reduced likelihood of success, as we tend to dip into the pool of funds for unexpected things. My recommendation here is to never dedicate more than 50% of your cash flow to your goal. That way, you still have funds to manage your day-to-day lifestyle.

Interest rates on borrowed money are not up to us as consumers to change, unless—as in my earlier example—we have an option to restructure. Looking at return on investment, if we plan on getting double-digit returns in a single-digit environment, we will likely be disappointed at best or potentially lose a significant portion of our investment at worst.

Ultimately, the options for most people are to either change the goal or change the time. What's the point in setting the goal if we don't keep it? Changing the time is probably the most prudent option.

In monitoring your cash flow and setting goals, it is recommended that you round up your expenses and round down your income. If you plan to have more expenses and less income, you will always be pleasantly surprised. Unfortunately, most of us do the opposite.

One of the primary reasons that anyone saves money and pays off debt is to maintain or raise his or her standard of living through each of the five levels of the Standard of Living Hill: dependence, independence, quality, comforts and luxuries. Once we have pushed the snowball up the hill, how do we keep it there?

Risk Management

Our standard of living is moved forward and held in place by the wedge of our income and assets. The strength and stability of the wedge is heavily influenced by the risks we take. Risk management is, in essence, the safety net we put in place. There is no crystal ball that will allow us to predict the future, but we can and we *must* plan or prepare for it. It is prudent to hope for the best while preparing for the worst. Ultimately, we can simplify risk management to three options: avoid the risk entirely, finance the risk or transfer the risk.

Option 1: *Avoid the risk*

If we are seriously concerned about the possibility (risk) of dying in a plane crash, we just never get onto a plane. Few risks, when it comes to personal finance, can be so easily avoided. Often, if we avoid one risk, we are faced with another.

As an investor, you may be concerned with the risk of losing money in stocks or mutual funds, so you avoid it by putting your money in guaranteed investments like GICs. But by avoiding market risk, you subject yourself to inflation and liquidity risk.

Inflation is guaranteed. In a year from now, the items you buy—like groceries, utilities and clothing—will cost more than they do today. Guaranteed. This is commonly referred to as the cost of living, and it is also known as inflation.

If you invest your money in a guaranteed investment that pays 2% and the cost of living goes up by 3% you just lost 1% because the value or purchasing power of your money went up by 2% but the items you want to buy are 3% more expensive. This ignores the possible tax consequences of receiving a 2% interest rate on your investment, which we will cover in more detail later.

The other primary risk with a guaranteed investment is liquidity. Most guaranteed investments are not redeemable. This means your money is locked away, and you cannot have it back or access any portion of it for a specific period of time that you agreed to when the money was initially invested.

You never know what the future may bring. If something happens where you need funds for a purchase or expense unexpectedly, you do not have access to these funds you have locked away. At this point, your choice is to miss the opportunity because your money is locked, liquidate other assets that you may prefer to hold on to or borrow and pay interest on borrowed money at least until your investment is up for maturity and you can access it. Now you have an understanding of the risk involved with avoiding risk.

Option 2: Finance the risk

The next option for risk management is to finance the risk, which in essence means setting aside additional money to cover or prepare for the possibility that you may need it. Knowing that GICs offer a lower rate of return, you set aside more money than what you need to offset the increase in cost of living and taxes on the interest earned.

But to be able to do this, you must think and plan ahead, and you will be sacrificing more of your current cash flow.

Option 3: Transfer the risk

The final option in our simplified risk-management model is to transfer the risk. This is done by having the appropriate protection in place in the form of insurance. This is a topic of its own and will be covered in another chapter.

Recap

To recap, we all have the choice to avoid risk, pay for it ourselves *when* (not if) it happens or protect our wedge by securing it in place at our current standard of living with the safety net of a protection strategy. The next step in your journey is to use the templates provided in this chapter to create a budget, a cash flow statement and a net worth statement.

Personal financial success is absolutely possible for everyone. By creating a plan and using these tools, you already have an ace in your hand at all times, which substantially improves your odds of winning this game.

3 Understanding Credit

Too many people spend money they don't have, to buy things they don't want, to impress people they don't like.
—*Will Rogers*
Actor

Now let's look at the next card in the deck. This is one area that most people do not fully appreciate: the importance of their credit rating.

TransUnion and Equifax

In Canada, there are two main companies that collect information from banks and finance companies related to how you handle your credit accounts: TransUnion and Equifax. I will summarize the most important points, but you should go to their websites and look around. Knowledge is power. The more you learn, the more you know, the more successful you will be at everything you do. As I am writing this, the websites are transunion.ca and equifax.ca. As we all know, websites have a habit of changing periodically (if memory serves me correctly, I believe TransUnion was tuc.ca at one time), so if those addresses don't work, just Google it.

TransUnion and Equifax collect information about you, including name, address, date of birth, social insurance number, employer and income. This information comes from the details you give and prove to the bank or finance company when applying for credit. Then, as you use and pay your loan, mortgage, line of credit and credit card, the bank reports back to TransUnion and Equifax how much you borrowed, what your credit limit is, how much the minimum payment is and whether you are making payments on time—meaning on or before the due date on the statement.

TransUnion and Equifax put this information into algorithms (mathematical equations) to create a report with two main credit scores which finance companies then use to determine their risk—probability of loss—in lending you more money. This report is called a *credit report* or *credit history*. Essentially, it is a report card of how you have managed the money you have borrowed in the past.

When a bank obtains your credit report, it receives a list of all your credit products. This includes details of things like your loans and lines of credit, as well as how you are managing your debt. This information is used to predict the probability that you will pay back the money you are asking to borrow. However, the importance of your credit doesn't stop there.

More reasons your credit matters

Many companies and businesses use your credit history as part of the job application process. Blemishes on your credit report can affect your ability to get a new job. Cellphone companies and utility companies also use your credit history to determine whether they will provide service to you. Likely they will not refuse to provide you a service, but depending on your credit, they may require that you make a significant deposit prior to giving you service. They will hold that deposit for the length of time they see fit to eliminate their risk. Even though many of these providers do not report your relationship to be part of the credit report, they still use your credit history as a tool to predict your likely future behaviour.

Due to the significant importance of your credit history, it is imperative that you contact both TransUnion and Equifax at least once or twice a year to verify that this information is correct and current. If the information is not accurate, you can request that it be corrected. You can ask for a copy of your credit report to be sent via Canada Post for free, or, for a nominal fee, you can go to the websites for TransUnion and Equifax and get it instantly along with your credit score, an explanation of what your score means and recommendations for how to improve your credit rating.

Doing this will also reduce the effect and negative impact of identity theft. If someone does get your information and applies for credit in your name, you will notice it early and be able to take the appropriate steps to correct it. Both companies also offer credit monitoring services which notify you when someone obtains a copy of your credit history.

The last thing you want to happen is to be applying for a mortgage for your dream home only to be told that you can't be approved due to a negative credit report—and you discover that the issue is related to a credit product that you never applied for. It could take months for the credit bureau issuers to investigate and resolve the matter, and the bank won't give you a mortgage until it has been fixed. Goodbye, dream home.

The Five Cs

Your credit history is not the only thing lenders look at when you apply for credit. A lender will also check what is referred to as the Five Cs of credit:

1. *Character*—How long has the applicant been at the current place of employment or in the same or similar role? How long at the same address? What is the reason for the loan? Is there a previous relationship with the lending institution?

2. *Capacity*—Does the applicant have the ability to repay the loan? This is straight math. If you subtract your expenses from your income, is there enough money left to make regular payments on the money you are asking to borrow?
3. *Credit*—What does the applicant's credit history look like?
4. *Collateral*—Does the applicant have assets that can be used as security?
5. *Capital*—What is the applicant's net worth? If the worst-case scenario came to pass and everything you own had to be sold, would there be enough money to pay off all your debts?

Generally speaking, if you meet four out of five of these, your application would be approved. If you have only two out of five, you are very likely not going to get what you are asking for.

For those of you wondering about three out of five, very simply, that is a grey area and is dependent on a very large number of variables. I will not try to analyze all of the possibilities here, but I will tell you that the most heavily weighted categories are typically *credit* and *capacity*. Some companies even go so far as to boil the traditional Five Cs down to three more-general categories:

1. *Ability*—Does the applicant have cash flow and the capacity to pay the money back?

2. *Stability*—How likely is it that the applicant's income will continue? Is this individual's job stable? Has he or she been at the same residence for a long time? Based on stage of life—age and maturity—is this person likely to pick up and disappear right after we lend the money?
3. *Willingness*—Based on credit history, does this individual have a demonstrated willingness to keep a commitment or promise to repay debt?

The next question people will often ask me is relative to determining capacity. What measuring stick is used?

There are actually two measuring sticks used, and while the percentages vary slightly from company to company, the following is the standard accepted by the Big 5:

- total debt service ratio (TDSR) = 40% or less
- gross debt service ratio (GDSR) = 32% or less.

GDSR is the percentage of your monthly income that goes to housing costs. This includes rent or mortgage payment, property taxes, heating and condominium fees. TDSR is the above plus all other debt payments, such as car loan or credit card payments (as reflected on your credit history) and spousal or child support payments.

If you have an income of $4,000 per month, pay $900 rent and have a heating bill of $100, your GDSR would be 25%. Add to that a $200 car-loan payment and a 3% minimum payment on $5,000 in revolving credit balances between credit cards and lines of credit ($150), and the math goes as follows:

$$\$900 + \$100 + \$200 + \$150 = \$1,350$$
$$\$1,350 \div \$4,000 = 33.75\% \text{ TDSR}.$$

You and your bank

While we are on the topic of borrowing money, let's talk about day-to-day banking. The majority of Canadians deal with two or three financial institutions for their chequing and savings accounts. My question is, why? There are many reasons you should pay fees for services with financial institutions that make sense, but why pay fees to several banks for the same thing?

A bank, when you clear away everything else, is a business. For a business to stay in operation, it must remain profitable. To receive the benefits of that business, you pay for the service you receive. Why pay multiple businesses for the same service?

I have seen many clients paying $300 to $500 a year in service fees for chequing and savings solutions at several different institutions when they could have been paying less than $150. The analogy I like to use is, if you are going to buy a jar of Kraft smooth peanut butter for $9.99 at one store and another store that is equally convenient to get to sells the same size for $6.99, why pay the higher price?

Granted, financial services are slightly different than good old PB. You need to also evaluate the quality of service and advice. Doing business with people who understand your situation, listen, care, give sound advice and are reasonably available to see you when you have the need pays far greater dividends and is worth paying a little extra for. Think about it. When you bought your last car or house, did you buy the cheapest one you could find?

Another pitfall I have seen is people failing to review their monthly bank statements. They have one of those accounts that offers ten or fifteen transactions a month for less than five dollars, but they are being charged a dollar per transaction over the allowable limit, and they are using the account actively for forty or fifty transactions. Having the wrong type of account can be very costly.

Go now and look at your most recent bank statement, or log on to your online banking. What were your service fees last month and the month before? Have you graduated to a bigger monthly plan in usage but still have the account that your parents set up for you when you were four years old? Did you realize that you could change your account type?

Considering the importance of your credit history, you should take some time right now and go to the websites of TransUnion and Equifax to get a copy of your credit report. Don't worry—getting a copy of your own report won't affect your credit rating. Make sure the information on that report is accurate. If your report says your credit history isn't very good, start working today to improve it.

One of the best ways you can improve your credit report is to pay your bills on time. The second thing I can recommend for improving your credit score is to bring your balances down. If you have a limit of $2,000 on a credit card, try to keep the balance below $700; that's 35% of the limit, which is a good ceiling to keep in mind when using your credit.

By knowing the Five Cs and knowing your credit history, you'll give yourself a fairly good idea of whether you will be approved for a loan before you even go to the bank. That is a very strong card to hold in your hand.

4. Credit Vehicles

I haven't reported my missing credit card to the police, because whoever stole it is spending less than my wife.
—*Ilie Nastase*
Tennis Player

THERE ARE TWO different vehicles you can use to borrow money: instalment vehicles and revolving vehicles. In order to be able to stack the deck when it comes to borrowing money, it is imperative that you understand the design of these vehicles so that you can use them properly.

Instalment vehicles

A loan is classified as an instalment vehicle. In its most basic form, it is borrowing money with a specific payment for a set period of time. The full sum is paid to you at the beginning, and to get more money at another point, you have to apply for a new loan with new terms and conditions. The variety of options is extensive and may depend on why you are borrowing the funds.

Loans may be secured or unsecured—the difference being that with a secured loan, your lender can take something from you if you don't pay the money back. A secured loan typically offers a lower interest rate, as the lender can recoup some of its lost money and interest income if you default on the payments by selling the asset that you offered as collateral.

A loan is a good place to start out with credit, as it creates the habit of owing money and paying it off entirely. As with many things, habits are developed over time, and usually bad habits are much easier to develop than good habits. Also, bad habits seem much harder to break than good ones.

Revolving vehicles

Credit cards and lines of credit are classified as revolving credit vehicles because you receive a specific limit and you use as much or as little of the limit as needed, when needed. The interest charged and the minimum payments required are relative to the balance owed. No balance means no payment is due and no interest is charged. You have the ability to pay and reuse with no specific point where the balance is required to be zero.

The best rule when it comes to credit cards is to pay off the entire balance you use each month before the due date. Interest rates on credit cards typically range between 20% and 30%, which is great for the bank but not so much for your financial plan. There are many credit cards that have rewards programs attached to them, which can give you great benefits if you are paying off the balance in full monthly. Once you start paying interest, the benefit is quickly reduced or eliminated.

Get paid every time you spend

Here's a perspective that many may disagree with, but that's okay. I firmly believe this is sound advice, as long as it's followed to the letter.

You have to buy stuff to support your lifestyle. There are only a few payment methods that you can use: cash, debit or credit. I have purposely ignored cheques, as most places no longer accept this as a viable payment method.

When you use cash or debit, the money leaves your possession immediately. You buy the item and it's yours. That's the trade-off. You acquire ownership of the purchased item in exchange for the cash because you feel that ownership of the item is worth more than the ownership of the cash. If you use a credit card with a reward, then you also receive that reward. I like cash, so I have a preference for a credit card that pays me cash rewards. Here's an example of why that's a good idea.

For a long time, my parents paid for everything with cash. It made their budget very simple. You get paid. You take the cash and spend it. When the money is gone, you wait to get paid before spending more money because you have to.

For a long time, they also had the belief that credit cards were used only by people who didn't have the cash to buy what they wanted. This may have been accurate in 1960, but things have changed a bit since then. The other thing they were concerned about was getting a big credit card statement at the end of the month that they might not be able to pay off in full, and then they would "get screwed by the bank with interest charges."

Using cash is great for people who live paycheque to paycheque. For them, there is always month left at the end of the money, meaning they run out of money before their next payday. This is likely caused because they don't like to budget. There's that "B" word again.

A few years ago, I sat down with my parents and explained my method: when you go to the store and you buy something, buy it with a credit card that pays you cash back. At the end of *every day*, use telephone banking or Internet banking to pay off the balance of the credit card from your bank account.

Now you aren't carrying around cash with the risk associated with losing it or having it stolen. When you receive the monthly statement, the balance is essentially zero, as you have been using a pay-as-you-go methodology.

You have all your transactions in one place so they're easier to track, and you are getting paid for all of your transactions (through the cash-back rewards). Cash doesn't do this. Debit doesn't do this.

My mother was comfortable with telephone banking, and my father was okay with Internet banking, so I sold them on the idea.

"Well, we'll give it a try, Kevin," my father said, "but I really don't think this will mean very much to us. We don't spend a lot, and paying it every day sounds like a lot of hassle for such a small benefit."

About a year and a half later, at Christmas, I went to visit them with my family. Within minutes of our arrival, my father looked at me and with a childlike grin said, "Come here, come here." He pulled me over to a quiet corner and said, "I'm not sure if your mother would mind me telling you this, but I have to say something. Actually, first off, I have to say thank you."

"For what?" I said.

"Christmas this year was free," he said.

"Should I be worried about the police showing up?" I asked with a smile.

"No, it's that cash-back credit card strategy you told us about last year. We got our cash back in November, and it was enough to buy all the Christmas presents. I can't believe that we got that kind of a reward. Wish I had thought of this years ago, thanks."

Wow.

To this day, they still do the same thing, except now that they are in the habit, I believe they only pay the card once a week instead of every day. Cash is used about as often as they previously used their credit card—only when they have to. And I never told my mother that my father told me about this. Well, maybe I did now. Hope you don't mind, Dad.

Mortgages

For simplicity's sake, a mortgage means borrowing money against the value of the home that you own or are purchasing in the form of a long-term loan. If you don't pay the loan, you lose the house. As with a secured loan, because the bank can take your home if you don't pay the loan back, you receive preferential interest rates.

Many financial institutions have home-equity plans where their lien on the house is, in essence, an umbrella under which you can have multiple credit vehicles. As long as your credit is good, you can move balances and limits around with relative ease. Oftentimes, as you make payments to your mortgage, the portion that is paying the principle goes back into the umbrella, and at any point in the future you can access the funds by adding it to another credit vehicle without going through the process of a credit application, employment and income verification—or this may be done automatically for you.

The benefit is a long-term borrowing solution to ensure that you always have access to borrowed money at the lowest available rates with the most flexible terms, and typically this is done with little or no expense to you. I think it goes without saying that this is a tremendous advantage to your financial plan.

As much as we want to avoid it, most of us, for most of our lives, will have to borrow money for various reasons. When we do, if we don't have a good rate in conjunction with a sound plan, we probably won't achieve the goals that we want to. If you are a homeowner, even if you have no balance owing, you should have an equity plan in place. There are so many circumstances in which people find themselves needing money that it is important to have a plan to manage risk and take advantage of opportunities that come up. This could be an opportunity to buy an asset or make an investment, or it could be a family or medical emergency. Either way, having quick access to money can be extremely important.

Don't wait until you *need* money to go see an advisor. That's not the best time. I can't say it often enough: you need flexibility, you need options. An equity plan gives that to you.

Now that you understand the different vehicles that are available to you for borrowing money, I suggest that you look at your situation. If you are not paying off your credit card every month in full, start doing that now. Look at your statements or online banking. Are the balances on your credit vehicles decreasing month after month? Are they increasing? If your balances are not decreasing month after month, maybe you should talk to your bank about changing your revolving credit into instalment credit through a consolidation loan or home-equity loan. By reorganizing your balances into lower-interest instalment vehicles, you immediately improve your cash flow, and over the long term you substantially improve your net worth.

If you want the deck to be stacked in your favour, you must be continuously increasing your cash flow and your net worth.

5 Savings and Debt Management

Do not save what is left after spending, but spend what is left after saving.
—*Warren Buffett*
CEO, Berkshire Hathaway

This is the simplest way I can put it: save something. Once you have started monitoring and tracking your income and expenses, and reorganized your high-interest debt to a lower-interest-rate vehicle with your new home-equity plan, you should now have increased your cash flow to the point where you can actually start saving.

There are many things to save for—including vacations, new vehicles and retirement—but by far, the first and foremost goal of saving should be to create an emergency cash reserve. The standard in the financial-planning world is to have three to six months of lifestyle expenses saved in an easy-to-access place, like a savings account. By having an emergency cash reserve, you should be able to circumvent additional borrowing in the future, relatively speaking. It makes more sense to save for a purchase in most cases than to finance it—especially when the purchased item is not going to improve your cash flow or net worth (due to no resale value or depreciation).

Certainly, it makes sense to have an emergency cash reserve. If something happens unexpectedly, like a disability or a family emergency, or you are not working for a period of time, you still need to pay your bills. If not, you may be in a position where you have to sell some of your assets in order to meet your lifestyle needs. Typically, three to six months is long enough to assess how this monkey wrench that has been thrown into your life is going to play out. If you do really need to sell assets, it gives you time.

Pause.

Time.

Time allows to you get a reasonable price for the asset you are selling, or for the situation to be resolved so that you are not forced to sell something at a time which may not be in your best interests in the long run.

Time

Time is arguably the most valuable asset we have, yet so many of us ignore it. If you spend money frivolously, you can do many things to get more money. Once time has been spent, you cannot get it back. Period. Whether it's time for your assets to increase in value or time with your family, lost time is gone. So if you can put a savings strategy in place that will buy you time to make an informed decision, don't ignore the opportunity.

I had a client a few years ago who was a perfect example of this. He and his wife had separated. When they did, his wife moved back to her home country and took their teenage son with her. It wasn't long after they got there that his son started to get into some trouble. The boy felt very isolated—he didn't really speak the language, he felt others his age treated him differently because he was of mixed race, and he quickly fell into depression.

When my client was talking with his son on the phone one evening, he became very concerned about his son's mental health, and he came to me the next day to talk about this. He told me he was thinking of taking some unpaid time off work to go see his son. Since we had a reserve of funds set aside, we calculated he could afford about four months off work. So he talked to his employer and made the arrangements.

My client ended up spending about two months visiting with his son, and when he came back he told me that his son confided in him that he had been contemplating suicide. The time spent with his father really helped him overcome the challenges he was having at adapting to the new living arrangements. Having a plan gave my client time to make a difference.

Pay yourself first

What is the most effective approach to getting this miracle fund of three to six months lifestyle expenses in place? The first time I heard of this concept was when I was in college and took a personal finance course. One of the required readings was *The Wealthy Barber* by David Chilton. The book's concept is very simple: save something.

Most financial-planning educational materials recommend that the sweet spot for savings is 10% of your income, each time you get paid (and there are differing opinions as to whether that's 10% of your gross income or 10% of your net income). But in fact, you are a unique individual. Your savings needs are different from any one else's. You need to determine what *your* savings needs are. Whatever you choose to put aside, the best strategy is to start early, start small and then work your way up.

You need to consider your financial plan as an obligation, like paying your cellphone bill or your rent. It can't be optional. Set up a savings account at your bank—this could be a regular savings account, a Tax-Free Savings Account (TFSA) or a Registered Retirement Savings Plan (RRSP), depending on your situation—and set up an automatic, recurring transfer from your chequing (transactional) account to your savings account.

If you get a pay raise, increase your savings. Review your savings strategy at least once a year and make sure it's still working. Create a plan and work on it until you have that emergency fund of three to six months lifestyle expenses saved. This gives you the safety net to help protect against the unexpected.

Start now.

Today.

Every time you get paid, set some money aside, even if it's only five dollars. Get in the habit. Once you've started, you'll notice that you don't miss it. Have your accounts set up so each time you get paid, you have the chosen amount automatically redirected to a savings plan. Then once or twice a year, re-evaluate and see if you can increase it by a small amount. This first savings plan should be your emergency fund.

It's worth the wait

Until you have your emergency fund in place, it's a good idea to have a line of credit to protect you from unexpected situations. Make sure, though, that the line of credit is not used for nice-to-haves. That is how people get into a very difficult situation of continually increasing debt—they sacrifice what they want most for what they want right now.

In our technology-driven society, we have come to a point of needing instant gratification for most things. Waiting for a Web page to download for more than thirty seconds often causes us to click, click, click and mumble about slow Internet service. Waiting months or years for anything has almost become passé.

Have you ever been in line at a store or in traffic and started complaining about the wait? Or perhaps become vocal, even angry, because things were moving too slow? Do you know anyone else who has? I suspect that everyone just said yes or nodded to that.

So it's no surprise that people have difficulty waiting to buy that big-screen TV that they want even though the one they have works fine. Maybe your neighbour went and bought a sixty-five-inch 4K Ultra HD LED Smart TV, and you feel this entitles you to one as well. Right? Well, just because the neighbour bought that sixty-five-inch TV or that brand-new SUV doesn't mean he could afford it. I've seen a number of people with six-figure incomes filing bankruptcy.

Once you have your emergency fund in place, start thinking about saving for the long term. Remember, it's your net worth that will determine how far your standard of living snowball slides down the hill of life when the income is no longer there to support that standard. Without assets, your SOL (standard of living) is SOL (shit outta luck).

6 Investing

October: This is one of the peculiarly dangerous months to speculate in stocks. The others are July, January, September, April, November, May, March, June, December, August and February.
　　　　　　　　　　　　　　　　　　　　　　　—Mark Twain
　　　　　　　　　　　　　　　　　　　　　　　　　　Author

Since I like to oversimplify things, here is how investing works: You give your financial institution your money that you don't need for your immediate lifestyle needs. They put that money into different types of investments, which are supposed to grow. *Supposed* to grow.

Yes, I said that twice on purpose. All investors have experienced occasions where that didn't seem to be the case, especially recently when there has been a lot of uncertainty in the global economy. However, when we look at history, the markets have always eventually recovered and gotten better.

I like the umbrella analogy that I started earlier, so I'm going to continue to use it for this as well. When you make investment, either yourself or through an advisor, you have three umbrellas under which they can invest that money:

1. *Registered Retirement Savings Plan* (RRSP)—This umbrella, discussed more in detail later, gives you a tax break for putting the money in. RRSPs grow tax-free while under this umbrella, but you pay

taxes when you take the money out—at the tax bracket you are in when you take it out.
2. *Tax-Free Savings Account* (TFSA)—Money goes under this umbrella without a tax break, grows tax-free and comes out tax-free.
3. *Non-Registered*—When your money is under this investment umbrella, you pay taxes on it every year, in most cases.

Under each of these umbrellas, you can invest money in virtually all the same things. This includes cash; term deposits or guaranteed investment certificates (GICs); mutual funds; stocks; bonds and many other vehicles. I am going to focus on just the first three: cash, GICs and mutual funds.

Cash and GICs

Cash is, well, *cash*. It is liquid, easy to access and represents no risk to the principle—and, sadly, no growth. In most cases, the interest that any company will pay to you for a cash account is nominal or non-existent. Why?

For exactly the same reasons you like it: you can take it and spend it at any time. When that happens, the financial institution or investment company doesn't have it anymore. Why is that a problem? It's your money, right? Yes. It is your money. But any financial institution that you deal with, for the honour of doing business with them, lends your money to other people. It's hard to do that if you are going to ask for it back the day after tomorrow.

Here's a banking secret that is important for you to know: a bank, or any company in the financial industry (including the government), is a business. Businesses need to make more money than they spend or they won't be in business anymore. You would be surprised at how many people seem shocked when I tell them that.

"Banks make billions of dollars in profit every quarter!" they say. "Why can't they charge us fewer fees?"

The answer, of course, is that fees are one of the reasons they are able to post profits and pay dividends to their shareholders every year. If you were a shareholder, you would want to have your shares increase in value because the business is successful, wouldn't you?

With a GIC (or term deposit, depending on who you deal with), you put your money into a vehicle and the bank locks the doors on your new ride, telling you they'll give you the keys back in a year, or two, or three or whatever. In return for your promise not to take the money away for a specific period of time, they will give you more interest than a cash savings accounts, and everything is guaranteed. It's a contract.

That is what many people consider to be the benefit of GICs—the guarantee and lack of uncertainty. Some people may even see the inaccessibility of the money as an advantage, as it ensures they can't spend the money. You say, "Here's my money. I promise not to touch it for a specific period of time." The financial institution says, "Thank you, and we promise that when you come back to get your money, it will still be here—and not only will we give you your money back, we also promise to pay you back this specific amount more." Everything is predetermined at the start of the agreement. No one can back out, in most cases.

Remember, this is my oversimplification. There are probably dozens of different flavours of GICs: stock-indexed, market-indexed, laddered, redeemable, non-redeemable. I am not trying to cover all of the chocolates and rocky roads. I am just giving you the simplified version.

Both cash and GICs (up to five-year lock-in terms) are protected by the Federal Government of Canada under the Canadian Deposit Insurance Corporation (CDIC). In the unlikely event that the bank goes bankrupt, CDIC will reimburse you the value of your account up to $100,000 per account. This is another layer of security and protection for the benefit of the Canadian consumer. Very nice. Thank you, Canadian government!

The challenge with GICs and cash is that the amount of interest you receive, or the rate at which the money grows, is typically less than inflation. Right now, today, you could go spend your $500 and buy stuff. Or you could invest it for a year, or two or five, in GICs. When you take the money out in one or two or five years and go to buy that same stuff, now that same stuff costs $520, but your investment only grew to $510. That's inflation deteriorating your purchasing power. Why would I save money if I can buy more stuff with it today than I will be able to tomorrow?

Don't get me wrong—I'm not saying don't invest in GICs. They can be an important aspect of a well-balanced portfolio due to their guarantees and predictability. But you need to understand the very real risk of loss of purchasing capacity when your money is in GICs. You have to be comfortable with that in order to make an informed decision on your investments.

Also, what happens if life changes, you need the money sooner than you thought and it is locked into a GIC for another six months or three years? What risks are you exposed to then? Remember, you can't take it out early. Most institutions do have "financial hardship" or "shortened life expectancy" exceptions incorporated into their GICs, but if you need that money and it isn't for one of those two reasons, you might not be able access it until the end of the term agreement.

In short, these are the advantages and disadvantages to GICs:

- *Advantages* include guarantees, predictability and the ability to keep your capital safe and inaccessible.
- *Disadvantages* include low rates, deteriorating purchasing power and that same inaccessibility.

Mutual funds

Essentially, a mutual fund is a pool of money collected from many investors and invested together as a single investment. Inside of that mutual fund there are three primary vehicles that are invested in: cash, fixed income and equities.

Cash, we understand and have already discussed.

Fixed-income refers primarily to bonds. With a bond, in essence you become the bank. You are the lender, and who is borrowing your money? Municipal, provincial and federal governments as well as corporations. Bonds are often very expensive for an individual investor, as it may cost hundreds of thousands or millions of dollars to buy an individual bond. A mutual fund allows people to put their money together and share in the interest paid by these businesses and governments to borrow the money from you. Bonds, like GICs, are usually issued for a specific term with fixed interest.

The other type of investment inside a mutual fund is stocks, also known as *equities*. With stocks, you are buying a piece of a company. You become an owner of the company when you buy its stock—a small owner, but an owner nonetheless. From a risk-versus-return perspective, stocks are near the top of the "risky" list for the average investor. As we have seen since the dawn of the market, individual companies can rise and collapse in a day. If you put all your eggs in one basket, you face the possibility that tomorrow the company won't be around anymore. Inside of a mutual fund, however, you are investing in the stock of many companies, not just one.

Most mutual funds maintain a mixture of stocks, bonds and cash. The mixture depends on the goal of the fund. Many mutual funds have minimum investment amounts as low as $500. For the average investor, this is the best option, as it gives a diversified portfolio of cash, bonds and stocks for a small initial investment. With $500 in a mutual fund, you could have stocks in all of the Big 5 banks in Canada, as well as government bonds and some Apple, Walmart and Loblaws stocks. If you tried to do this outside of a mutual fund … well, a single share Apple stock alone is over $100 (as of this writing).

In order for you to lose all your money in this hypothetical mutual fund, Apple, Walmart, Loblaws, the five biggest banks in Canada and the Canadian government would essentially have to file bankruptcy simultaneously. I would suggest that the odds of that are phenomenally low, and if it did happen, we would have a much bigger problem than where our money is.

With a mutual fund, you have built-in diversification by nature of design. The other advantage of a mutual fund versus guaranteed investments is accessibility. The majority of mutual funds can be accessed at any time without penalty, except during the first thirty days.

To recap, here are the advantages and disadvantages of mutual funds:

- *Advantages* include diversification, accessibility, potential for higher returns and affordability.
- *Disadvantages* include the potential for loss of money.

Trade-offs

Everything is about trade-off. When you are playing poker, you can sit on the cards you have or take a chance and trade in a couple of cards to try to get a better hand. Which cards you trade in and when are entirely up to you.

When you invest in cash or GICs, you are trading purchasing power and accessibility in return for security and guarantees, and there are risks associated with that. When you invest in mutual funds, you are trading in security and guarantees for accessibility and potential return, and there are risks associated with that.

Personally, I like flexibility. I like to have options. If I invest in a mutual fund, I can change my mind about the mix of investments by moving to a more conservative (more bonds and cash) style or to a more aggressive (more stocks) style, if I decide that it is a good idea—of course, with the guidance of my trusted advisor. When it comes to GICs, I know my advisor will tell me that I can't change my mind or access the money until maturity unless I have a redeemable certificate, which is probably paying such a low rate that I might as well have the money in a savings account. If we are ever in a situation where I can get 4% or more on a GIC, then I may be more inclined to put more there, because at 4% I can at least keep up with inflation and some of the taxes.

Understand your investments

Regardless of which decisions you make in life, one way to stack the deck in any area is to find someone who has been successful doing it in the past and do a bit of R&D. No, not research and development: rob and duplicate. When it comes to investing, I would suggest that it makes sense to follow the advice of someone who has been very successful at investing.

Warren Buffett is arguably the foremost investor of our time. If you don't know his story, you should look it up. The simplified version, as I understand it, is that he started with a relatively small amount of his own money and over time built Berkshire Hathaway into one of the largest investment companies on the planet. He has spent a few decades at or near the top of the *Forbes* list of the world's richest people. Sounds like a guy who might have some words of wisdom when it comes to investing.

Mr. Buffett has been known to say that he only buys investments he understands. So to follow his lead, if you don't understand what would affect the success of a telecom business in Asia, perhaps investing in a start-up company in that industry isn't a great place to hang your hat. When you understand the way something works, you're less likely to be emotional about it.

One very real challenge for investors is that it is easy to answer an investor profile questionnaire and say that you are comfortable with a 15% fluctuation in any given year, but when your $10,000 drops to $8,500, you may realize that you really don't have the stomach for it, because you don't understand the investment or haven't properly evaluated your goal or timeline.

I have included an investment profile questionnaire in the "Know Your Financial Self" section at the end of this book. Grab a piece of paper and a writing tool and go complete that questionnaire now. See how you do. This questionnaire will help you understand what type of investor you are: Conservative, Moderate, Balanced, Growth or Aggressive. You do not need to know anything about investing to do this; you just have to know yourself.

Buy high, sell low. That's the reality of many individual investors' approach. They see an investment go up day after day, week after week, month after month, and they hear everyone talking about buying in. They finally buy in near the top, they enjoy that excitement as the roller coaster inches up the last few feet of track, and then they scream like elementary school kids when the coaster crests at the peak and hurls them forward on a downward spiral. Probably somewhere between the first corkscrew corner and the next uphill, they have already jumped off the track and moved back into cash or GICs ... until the next time they are overcome with the greed that got them into trouble in the first place, when they hear their colleagues and friends talking about how much they made with a particular investment. They jump back in, and the cycle continues. Then they talk about how investing doesn't work, as they always lose money. The best way to avoid this is to understand your investments and stack your deck with strategies that take advantage of the knowledge you have.

Dollar-cost averaging

Once you have your emergency fund intact, you can convert your "pay yourself first" automatic savings plan into an automatic investment plan to take advantage of volatility in the market. You work hard for your money, and so your money should work hard for you.

Let's say you are saving $200 per month, and let's assume that you are putting your money into a mutual fund. When you buy a mutual fund, you are buying units. In June, when you invest your $200, let's say the units are worth $10 each. In this case, you buy twenty units. In July, the market drops substantially, and your units are worth $5. Your $200 would then buy forty units. In August, the market goes back up to $10 per unit, and, again your automatic plan buys twenty units. At this point, you have eighty units with a value of $10 each, which is $800. You have invested $600, and therefore you have a $200 profit.

By purchasing more of an investment when the price is low and less when the price is high, you are lowering the average cost of each unit. This means you are getting more value out of each dollar you spend. Think of it like going to the grocery store and one of your favourite items, that you buy regularly, is on sale. You'll buy extra so that you can buy less when it's not.

The power of investing early

Regardless of what age and stage of life you are at, starting to save and invest now makes more sense than waiting until later. As I mentioned before, time is one of the most valuable commodities there is, so we must take advantage of the time that we have.

Let's have a look at two friends, Patrick and Anne. They are both 25, work decent jobs and have some surplus cash flow. Anne decides she is going to invest $250 every payday into an investment plan. Patrick feels he's too young to worry about retirement and planning, so he just spends his money having fun.

Fast-forward ten years. Anne and Patrick are still working, but now they are both married and starting a family. Patrick is realizing that he isn't getting any younger, and he decides he needs to start setting aside some money. After talking to his friend Anne, he decides $250 every payday should be affordable. At the same time, Anne has had some unexpected things happen in her life, and she stops contributing to her investment plan.

Life moves on, and they both turn 60. Patrick has been diligently contributing his $250 from age 35 until 60. Anne did a great job from 25 to 35 but was never really able to make any additional contributions between 35 and 60. In all, Anne contributed $65,000 of her own money—$250 every two weeks for the ten years from age 25 to 35—and Patrick contributed $162,500 of his own money—$250 every two weeks for the twenty-five years from age 35 to 60. If we assume that they both got the exact same annual return of 7%, at the age of 60 Patrick has $424,792 and Anne has $503,633. This is the power of time.

Diversification

If you do a bit of research, you will discover that a large portion of an investment return is asset allocation, which is how your money is invested. The concept of diversification essentially means to have a variety of different investments, stocks, bonds, real estate and guaranteed investments, which reduces the volatility of your overall investments.

When we are looking at total retirement income, one of those sources of income we discussed earlier is the Canada Pension Plan. As of right now, the CPP pays out an annual maximum of $12,780. Granted, most Canadians don't get the maximum—when you're doing your own calculations, you are probably better off using $6,000 or $7,000 as your estimated income from that source—but let's just say for this example that you are one of the lucky few who gets $12,000 from your CPP.

I look at the CPP as being like a GIC, generating guaranteed income taxable at my highest tax bracket each year so long as I am alive. For me to get $12,000 income from a GIC, I would have to have a $400,000 GIC at 3% or an $800,000 GIC at 1.5% (which is closer to today's GIC rates). If you already have an investment like this, do you need to buy more of them? Maybe you do. Maybe you don't.

At minimum, you should take these things into consideration when building your plan. I don't have a choice in how the CPP is invested, but I do have a choice in how my money is invested. Maybe I should consider these things when making investment decisions.

Building a $100,000 investment portfolio

To build a $100,000 investment portfolio, you could save $600 per month and eventually, over time, someday, build to $100,000. Or you could borrow $100,000 from the bank and pay the bank back $600 per month. Either way, it is $600 per month. The difference is that by borrowing $100,000, you have a larger upfront investment to take advantage of compound growth, which would have a larger impact on increasing your net worth over time versus a regular investment plan of $600 per month, assuming the same rate of return and time invested.

The best way for you to stack the deck with investments is to eliminate as much of the noise and emotion as you can. Set up an automatic investment plan in one or more of the available umbrellas: Tax-Free Savings, Registered Retirement Savings, or Non-Registered, depending on your needs and cash flow. Make sure you have specific goals in mind for each of your investments in order to help you stay on track. Review and monitor your plan at least once a year. Understand your investments and use strategies that take advantage of the knowledge you now have.

Finally, don't wait. Time is a valuable commodity when it comes to investing. You will be successful in investing when you start today.

7 RRSPs and Retirement

The question isn't at what age I want to retire, it's at what income.

—*George Foreman*

Boxer

Registered Retirement Savings Plans, also known as RRSPs, were introduced in Canada in 1957. The idea is fairly simple: if you put money into an RRSP, the government will return the taxes you paid on that income when you earned it.

Many people complain about how much of their paycheque they lose to taxes every two weeks. Well, this is your chance to get that money back. An RRSP contribution will reduce how much you owe at tax time, or it will increase the amount of your refund. How much depends on your tax bracket.

If you want to know specific details about your situation, have a look at the website addresses I've included in the back of this book for the Ernst and Young calculators. They are great tools for understanding how much tax you pay and what kind of reward you could get for making an RRSP contribution based on your income level. As with anything, the numbers generated on these sites are approximations and cannot take into consideration your unique situation—like if you have other allowable deductions or credits on your taxes, such as medical expenses or spousal support payments.

Getting money back sounds great, right? What's the catch? For starters, you can only contribute up to 18% of your annual income, minus any amount your employer puts into a pension plan, if it offers one. If you overcontribute to an RRSP, you are penalized 1% per month on the overcontributed amount. Technically, you can overcontribute up to $2,000 and not get charged the penalty, but I am not going into detail on that. Check the Canada Revenue website for more information regarding overcontribution to your RRSP.

The other catch is that when you withdraw the money from the RRSP, you have to include the withdrawn amount in your annual income for the year you took out the money. If you don't contribute the full 18% each year (including the employer pension amount), that amount goes into a pool of unused RRSP contribution room and carries forward into the next tax year. Each year that you don't maximize your RRSP, the contribution carries into the next year, pretty much indefinitely.

After you file your income taxes, Canada Revenue sends you a Notice of Assessment (NOA), which essentially confirms that it agrees (or disagrees) with the information you put on your tax return. On your NOA, you can see your total unused carry-forward RRSP contribution-pool value, which for many Canadians in their forties is likely to be $50,000 or $100,000, as most people do not use their full contribution room each year.

So really, an RRSP is a tax-deferral tool—based on the assumption that when we retire, we will be in a lower tax bracket in retirement than when we are working full time. For example, let's say you earn $75,000 and your marginal tax rate (the amount of tax you pay on the next dollar you earn) is 32%; if you make a $5,000 contribution to an RRSP, you would get about a $1,600 refund on your taxes, which is about the same amount your employer would have taken off your paycheque and sent to the government on your behalf when you earned it.

Fast-forward a couple of decades, and now you are retired, and your retirement income is $35,000. When you withdraw that same $5,000, your income becomes $40,000. At $40,000, your marginal tax rate (the rate you owe on that $5,000) is 20%, so you would owe $1,000 in taxes.

When you made the RRSP contribution, the government gave you $1,600. When you took that same amount of money out of the RRSP, you had to give back $1,000. You just made $600 by putting money into an account and withdrawing it. This is almost like magic, creating something from nothing.

For people in the 50% tax bracket, I look at it this way: you put $10,000 into an RRSP, and you receive a refund of $5,000. Now you have an investment plan with $5,000 of your money and $5,000 of the government's money. Over the years, the money is invested, and it grows tax-free. When it is time to take the money out, you take out your half and the government takes out its half. It's like an interest-free, payment-free loan from the government. Why wouldn't you want it? The only thing the government wants is for you to manage its money well, because it shares if your money goes up and it shares if your money goes down.

Once you have placed your money under the RRSP umbrella, you can—and in my opinion, you should—invest that money. As explained in the previous chapter, you can invest your RRSP in savings, GICs, mutual funds, whatever. The whole point of saving taxes is not just to save taxes but also to increase your net worth. If you are saving taxes but the investment itself isn't doing anything for you, then perhaps there are better ways to invest your money.

RIF, LLP and HBP

At some point, the government wants you to pay taxes on this money. As it stands right now, you may convert your Retirement Savings Plan into a Retirement Income Fund (RIF) at any age, but no later than December 31 of the year in which you turn 71. The year the money is converted to a RIF, no money is required to be withdrawn, but each subsequent year you are required to withdraw a minimum percentage of the total value. There is a chart on the Canada Revenue website which details the percentages based on your age.

There are also a couple of special programs involving RRSPs which allow you to withdraw your money without paying taxes to Canada Revenue. They are the Lifelong Learning Plan (LLP) and the Home Buyer's Plan (HBP).

The Lifelong Learning Plan allows you to withdraw from your RRSP to pay for full-time training or education for you and your spouse or common-law partner. This is not for your children; there is a different program for that called the Registered Education Savings Plan (RESP).

For both the LLP and the HBP, the money must be in your RRSP for at least ninety days prior to being eligible for tax-sheltered withdrawals. With the LLP, you can withdraw a maximum of $10,000 per year and $20,000 in total. Both you and your spouse can participate for yourself or each other. This means that if you and your partner each have $20,000 in your RRSP, each of you could withdraw $20,000, for a total of $40,000 for one of you to go back to school. Generally speaking, you have to pay back 10% each year into your RRSP under the LLP. It's like an interest-free loan from you to you (or your spouse)—which is better than borrowing the money from the bank.

For more information on the LLP, please check out the Canada Revenue website; it is a great source of information, and overall it's written in a fairly easy-to-understand way. Since you would have to withdraw from your RRSP to participate in this program, you would also have to consult your financial advisor, who can give you some guidance in this matter.

In my opinion, the most common reason people start investing in RRSPs is because they hear about the HBP as a great way to save for the down payment on their first home. The program allows you to withdraw up to $25,000 in a calendar year from your RRSP to buy or build a home that you intend to live in. In order to qualify, you must be a first-time home buyer, meaning you did not live in a home that you or your current spouse or common-law partner owned in the last four years. So, if you owned a home in the past, you may still qualify for the HBP. Your financial advisor can help you with the forms to make this determination.

Now, Canada Revenue understands that when you buy a home, during the first couple of years there are additional expenses—new furniture, a coat of paint, new drapes, etc. If you have been renting, your rent may have been inclusive, so you may not be accustomed to paying water or heat, and certainly not property taxes. In order to help you get acclimatized to your new financial obligations, Canada Revenue allows you to skip the first two years of putting money back into your RRSP to repay the HBP withdrawal. Again, this is like an interest-free loan from yourself to yourself, but it isn't payment-free; you do need to pay yourself back by recontributing whatever you withdraw over the course of fifteen years.

Let's say you withdraw $15,000, to keep the math simple for me. The first two years after your withdrawal, you don't have to pay anything back into your RRSP if you don't want to. After that two-year grace period, you have 15 years to put $15,000 back into your RRSP, or $1,000 per year. If you don't, then Canada Revenue will add the $1,000 each year to your income, and you will have to pay taxes on it when you file your income tax. Conversely, there is nothing holding you back from repaying that $15,000 into your RRSP faster. There is no minimum repayment period, only a maximum.

Spousal RRSP

The only other type of RRSP that I am going to cover is the spousal RRSP. This can have some great advantages, but like everything else, there are risks associated with it.

A spousal RRSP is an arrangement where you can put money up to your contribution limit, including your unused pool, into an RRSP in your spouse's name. When you do this, as the contributor, you get the tax refund, and when your spouse (the annuitant) withdraws the money, presumably in retirement, he or she pays the taxes. This is a good strategy for people for whom there will be a big difference in retirement income—for example, if one partner doesn't have a company pension plan.

My ex-wife stayed at home to raise our children for most of my career. As a result, she never contributed to the Canada Pension Plan. If she does go to work in the future and the company she works for offers a pension plan, she would have a reduced number years in the workforce, and therefore she will be receiving a lower pension and reduced CPP payments. If I work in banking for the rest of my career, I will have a pretty good company pension, and probably a fairly high CPP payment, and therefore we will probably be in different tax brackets. If I'm in a 40% tax bracket at that time, I lose 40% of my RRSPs to taxes when I withdraw them, and since some government income in retirement is geared to income, I may even have payments like Old Age Security (OAS) reduced or even completely clawed back by Canada Revenue when I start taking my RRSPs out.

If I plan ahead and see that coming, then maybe I should consider a spousal RRSP, because in retirement she will probably be in a 20% tax bracket. In the above scenario, when she takes money out of the RRSP she's in a low enough tax bracket that her OAS probably would not be clawed back *and* we just made $200 in savings on every $1,000 that we withdraw because it's taxed at her 20% not my 40%. As a reminder, when I put the money into the RRSP, we got the tax refund at my marginal tax rate. That really starts to make a difference when you are talking about $10,000 or $100,000 or $1,000,000 in an RRSP.

Where's the drawback? What's the risk? If money is withdrawn from a spousal RRSP in fewer than three years from the time the contribution was made, the withdrawal is taxable income to the contributor. Another risk is, if you want to use a spousal RRSP for the HBP, the contributor cannot take *any* money out; only the annuitant can.

An additional risk is if you and your spouse aren't dancing together forever, in most cases, the spousal RRSP belongs to the person who is going to take the money out, not the person who put the money in. In my case, she stayed at home to raise and home-school our five kids, sacrificed a career in forensic accounting and put up with me for 20-plus years, so to my mind she earned it (and probably more than that). It's hers, and I am okay with that.

Your situation might be different. You might look at it differently or feel differently. You just have to realize that when you put money into a spousal RRSP, it's not yours anymore. If your relationship works out better than mine, you have no concern, and this strategy will definitely, absolutely increase your total household income in retirement.

Planning for retirement

So many times I have had people come to me after they went to a seminar at their place of work because they are five years from retirement. They come to me and say, "Oh my God, I'm retiring in five years, and I don't know if I have enough money."

My initial thought at that point is: "Hmm, too late."

If you haven't started retirement planning by that point and you aren't (by sheer luck or benefit of an inheritance) already on track, you've waited too long. The discussion is more about which aspects of your retirement dreams you are willing to sacrifice. Work longer? Less travelling? Die sooner? Usually people aren't keen on the last option, so we have to work with the other choices, and reducing your lifestyle during retirement seems like defeating the purpose of working your whole adult life.

But what is that magic number? How much do we really *need* to save? If we save everything today, we sacrifice our standard of living because we aren't spending to enjoy experiences—and life is about experiences, both before and during retirement. There is a balance.

The goal is to understand the purpose of saving for the long term. You do it so that your savings and investments can replace your income when you aren't working for the rest of your life. This is often called *passive income*, which is income that you receive without going to work because your assets are working for you. You work hard for your money, and so your money should work hard for you. When your passive income is enough to support all of your lifestyle expenses, you have achieved financial freedom.

Financial freedom number

Once you know your FFN, you can work toward achieving it. The simplest calculation is to say that if you need an income of $40,000 per year for 25 years, assuming retirement is from age 65 to 90 (ignoring inflation, taxes and return on investment), you need $1,000,000 in investments: $40,000 × 25. Simple.

The more complex calculations come into play when you start considering all of your income sources. This could include employment and government pensions, RRSPs,

CPP and OAS, rental income, business income ... the list can be very lengthy. This is more complex for most people, because they have no idea what their sources of income will be let alone how much they will get from those sources.

The most important thing is to start by determining, "How much money do I need to have per month or per year to enjoy the lifestyle I want?" The best way to do that is to go back to the earlier chapter on the evil "B" word. The vast majority of people I have met over the years have no idea how much is enough. If you don't know what your goal is, you most certainly won't achieve it. Most financial reading materials will tell you that you need 70% of your current income to fund your retirement. I say that depends on the lifestyle you want. For some, it's more what lifestyle can you afford, because of the decisions you made before you started to plan.

Do you like the lifestyle you have now? Do you want to have this lifestyle for the rest of your life? These are good questions to start with, because they will help you measure your future needs and can help create a starting point by telling you what your FFN is. Your FFN gives you a tangible target to work towards. Once you achieve that goal, you then have the choice of whether you want to continue regular employment. That sounds nice, doesn't it? You have the *choice* of whether you *want* to work or not.

Retirement is the longest vacation you will ever take. So many people spend weeks or months planning their two-week summer vacation. Why don't they spend any time planning their retirement? Once you have determined your FFN, you are on track to planning your retirement lifestyle. The next step is understanding what your sources of income are during retirement.

Understanding retirement income sources

Thankfully, most of us don't have to rely on our own wealth to manage our retirement lifestyle. If we did, we would see a much larger number of people spending their golden years working at the Golden Arches. In addition to our own savings and investments, we also have some or all of the following:

- CPP
- OAS
- Guaranteed Income Supplement (GIS)
- RRSPs
- company pensions

As part of your planning process, you should go to servicecanada.gc.ca and click on "My Service Canada Account." This will allow you to access information concerning your CPP contributions and your likely CPP income via your Statement of Contributions. It is a good idea to review this information at least annually for accuracy, just like you should check your credit history for accuracy. If there is a mistake on your Statement of Contributions that will affect your future income, it will likely be difficult to remember specifics like your annual income fifteen years ago to be able to dispute any discrepancies. If you are checking this regularly, you can catch problems early when it is a lot easier to provide proof of the error.

Once you have your information from your pension-plan provider, the details from your Statement of Contributions and knowledge of your other income sources, we go back to the triangle from Chapter 2. This will tell you how much more you need to save to get to your FFN. The reverse works as well if you are already at retirement and have already done your savings, and you want to know how long the money will last.

RRSP strategies

Since you understand the benefit of dollar-cost averaging, you set up a regular investment plan into an RRSP for a total of $5,000 per year. When you file your taxes, let's say you're in the top marginal tax bracket, and you get a refund of $2,500. If you go to the bank, you may qualify for an RRSP loan for $5,000, and you put that in your RRSP. Now you have a $10,000 RRSP and a $5,000 refund. Most banks offer payment deferrals on RRSP loans, so you can

STACKING THE DECK

delay your first loan payment for three months or more. When you get your $5,000 tax refund, you pay off the RRSP loan, and you never had to make a payment on the loan. This strategy allows you to double your RRSPs every year without any additional cost to yourself.

A client came to me one time and told me that he wanted to do a $5,000 RRSP contribution. I asked him why. He told me it was because he wanted to pay less tax. Considering he was making $150,000 a year and was paying $48,000 in tax, I can understand why he wanted to pay less. I also knew he had a car loan that was down to about $15,000, and he was paying about $600 a month. Rather than make a $5,000 RRSP contribution, what we did was take out an RRSP loan for $35,000, which gave him a $15,000 tax refund which we used to pay off the car loan. We used the same $600 monthly payment to pay the RRSP loan instead, which had a better interest rate, and the RRSP was an asset which would appreciate, while the car was an asset that depreciates. He would have monthly payments for debt either way, and it seemed to me this was a better option.

Really, it's all about cash flow, and cash flow is king. Do you have enough cash flow being generated without you doing anything in order to not have to go to work every day? Retirement can happen at any age for anyone. It's about where that wedge gets put in place on the Standard of Living Hill. The lifestyle you live today and the lifestyle you want to live tomorrow are the driving factors behind how long you will continue to abuse the snooze button each morning because you really don't want to go to work but you have to.

This is your call to action. Go back and review the section about your FFN. Calculate your number, and then go to your financial institution, set up an RRSP and start your path to financial freedom. Visualize it. Believe it. You are on your way, and you are getting closer with every step you take.

8 Tax-Free Savings and RESPs

*How many millionaires do you know who have
become wealthy investing in savings accounts?*
—*Robert G. Allen*
Author

The Tax-Free Savings Account (TFSA) is a fantastic new tool that many people are taking advantage of. Unfortunately, too many have fallen prey to the misunderstanding created by its name. In my experience, many people are putting money into this amazing vehicle and just parking it there in a savings account or in cash.

The TFSA, in my opinion, would be better described as a TFIA: Tax-Free Investment Account. You can, and you should, be investing this money, because it grows *tax-free*. How often do you have an opportunity like that? Take advantage of it. You know it's a good thing if the government puts a limit on it. So use it.

The TFSA was introduced by the federal government in January 2009. All Canadians who were 18 years of age before the end of the calendar year could contribute to this new type of savings program. In some provinces and territories, the legal age to enter a contract is 19, and opening a TFSA is a contract, so in those provinces and territories a TFSA cannot be opened until 19 years of age. But the contribution room from age 18 carries forward for all Canadians regardless, and even if you haven't opened a Tax-Free Account. This means that if you don't contribute in one year, any subsequent year you can contribute any amount up to the maximum of all previous years that you were eligible but did not contribute the maximum, just like RRSPs.

For example, if you turned 18 in 2009, assuming you did not put any funds into a Tax-Free Account in 2009 or 2010, as of January 1, 2011, you could contribute into the TFSA the maximum of $5,000 for 2009, 2010 and 2011, for a total of $15,000.

This contribution limit is connected to your Social Insurance Number and is a combined limit regardless of how many financial institutions you deal with. It is your responsibility to ensure that you do not overcontribute. Excess contributions carry a penalty of 1% per month. The fun part is that the government doesn't know that you've overcontributed until the banks submit their information about how much you contributed. This is likely many months later. If you aren't paying attention, you could end up paying quite a lot.

For 2012, the government placed a contribution limit of $5,000. For 2103 and 2014, they increased it to $5,500. In 2015, the government increased the tax-free contribution limit to $10,000, and in 2016 it was reduced to $5,500 per year going forward … or, well, until they change their mind. That said, as of the time this is being written, if you were a Canadian resident and 18 years of age in 2009, your total tax-free contribution limit is $46,500. If you are a temporary resident and your Social Insurance Number starts with a 9, you may want to consult with Canada Revenue to verify your eligibility for this type of investment plan.

If you decide for some reason to deal with more than one financial institution, they will not know your business with another company unless you disclose it to them. This makes it more challenging to receive sound financial advice, so it is important to either tell them or deal with one company.

When you make an investment into a Tax-Free Account, the funds grow and can be withdrawn without tax consequences. However, unlike RRSPs, there is no tax benefit to making the contribution. Also, unlike an RRSP, if you make a withdrawal from a TFSA, you *regain your contribution room* the year after you complete the withdrawal.

For example, my son made a contribution in 2009 for $1,000, leaving $4,000 of contribution room unused. In 2010, he has the $5,000 allowable contribution room, plus the $4,000 he didn't use during the previous year, for a total of $9,000. During the year, he decides to withdraw $500 of the $1,000 that he put into the TFSA; now he has the $9,000 unused contribution room plus the $500 that he withdrew for a total of $9,500. But putting the $500 that he withdrew back into the TFSA must be done in a different year. Recontributing in the same year could create an overcontribution, which would be subject to the 1% penalty.

Any of the income earned or growth within the TFSA does not affect Old Age Security, Guaranteed Income Supplement or Child Tax Benefits. A person doesn't need to have employment income to make contributions—gifted funds can be placed in a TFSA—and assets in a TFSA can generally be transferred to a spouse or common-law partner upon death. Individuals named as "successor holders" can even choose to place the money in their TFSA regardless of how much contribution room they have. This is a great way to pass your legacy to your significant other without worrying about estate taxes.

This is the perfect medium- to long-term savings vehicle in many cases. If you have short-term savings needs, use regular non-registered savings/investment accounts. The money won't be there long enough to get taxed anyway. Go to the Canada Revenue website, find out what your TFSA contribution limit is, and once you have your short-term emergency fund in place, develop a plan to maximize the value and benefit of this amazing Tax-Free Investment tool.

For a long time, people asked, "Contribute to my RRSP or pay down my mortgage ... which is better?" Now people are asking, "Contribute to a TFSA or an RRSP ... which is better?" There is no one-size-fits-all answer for this question. Ask your financial advisor to help you determine the best strategy for you. The general rule of thumb is, if you expect your income to be higher later in life, put your money into a TFSA. If your income will be lower later in life, make investments in an RRSP.

And of course, considering the TFSA limit is low for some people, once you have maximized the TFSA, fall back on the RRSP option. If you have maximized your TFSA and your RRSPs, come see me so that I can switch your investments to my company.

That's it for the TFSA. Simple.

RESP

The last of the registered or tax-sheltered investment vehicles that I am going to discuss is the Registered Education Savings Plan (RESP). As with the TFSA, you as the contributor do not get a tax refund for putting money into an RESP. However, what you do get is a federal government match of 20% of your contribution to a maximum of $500 per year in free government money.

As with everything, there are a variety of ways an RESP can be set up. There is also a lot of bank jargon used when you do any research on RESPs, but these are the basics. You can set up an RESP for your children after they are born and you have gotten a Social Insurance Number for them. As of 2007, there isn't a limit to the amount you can put into an RESP in any given year, but for any one child the maximum lifetime contribution is $50,000. If Grandma and Grandpa are putting money into an RESP for the grandkids also, make sure you know about it, because if the child's SIN is flagged as having too much RESP money, anyone who has a plan set up for that child is taxed a penalty of 1% per month on any amount over the limit.

Once your child's RESP is set up, to take maximum advantage of the government's free money, you would put $2,500 per year into the RESP and the government would add $500. If you only put $1,000 in, then the government would match by contributing $200. Regardless of how small or how large your contribution is, the government will always match $0.20 for every $1.00 you put in *per child*, up to that $500 per year maximum.

If you happen to set up an RESP later in the child's life, you can get the government match on a maximum of $5,000 per year to catch up on the previous year's contribution limits. For example, if you set up an RESP when your child is 5 years old, by the time he or she is 8 years old you could catch up on the maximum contribution limit and maximum government grant match allotment by using this method.

What happens to the money while it is in the RESP? Hopefully, it grows if you have had some good advice about investing it, and the investment options are the same as discussed earlier about Non-Registered, RRSPs and TFSAs. You can continue to contribute to an RESP and the government will continue to match your contributions until December 31 of the year that the child turns 17. (Note that there is some fine print about this which you will either want to research online or talk to an advisor about when you are setting it up.) In my opinion, in most cases, this is when the free money stops being added, and this is when the major value of the RESP is lost. After December 31 of the year your child turns 17, refocus on your TFSA if you still have contribution room. You can take money out of the TFSA for any reason, including helping your kids with their education costs.

Technically, the money in an RESP can be withdrawn at any time, but if it is not for educational purposes, you lose the government's 20% on the amount you withdrew. But when the money is taken out for educational purposes as Educational Assistance Payments, you keep the government grant and the money withdrawn is taxable income for the student. If your child doesn't continue his or her education after high school, you can leave the money in the RESP for up to 35 years after the plan was opened and see if the child has a change of mind, or you can close it and take your money back, at which time the government also takes its money back.

An RESP is a fantastic way to get free money from the government to help fund your child's postsecondary education. By using RESPs to their maximum benefit, you can help your kids stack their own deck by helping them invest in knowledge without the weight of student debt.

9 Estate Planning

It's not about how much money you make, but how much money you keep, how hard it works for you and how many generations you keep it for.
—*Robert Kiyosaki*
Author

ESTATE PLANNING IS arguably the most ignored aspect of financial planning. Why? Because at that point you're dead, and a lot of people don't like to talk or think about this part of the life cycle.

We are born. We live. We die. That is the reality of the human condition. Not planning for it doesn't make it not happen (forgive the multiple negatives).

The individual components of your estate planning consist largely of creating a will, naming a power of attorney, nominating an executor and planning your funeral. More complex estates may also need business succession planning. I look at that as part of business planning and therefore will not discuss that in this book.

Your Will

A will is a list of all your stuff and who you want to have it when you're gone. This is important to reduce the likelihood that people will fight over your stuff. It is also important to ensure that your assets are distributed in the way you want. Lack of a will creates a lot of legal costs for your family because most people won't agree on who gets what. Even if it doesn't cause a full-scale legal battle, it could create difficulties in relationships because some people feel that others got what they felt they deserved or wanted from your estate.

In order to create a will, I strongly recommend going to a lawyer. There are home-kit versions available, but if your will is not written exactly as it should be, there may be room for someone to dispute it in court. A will written by a lawyer is more likely to be deemed valid and would be less likely to be disputed.

Power of Attorney

A power of attorney is a document that identifies one or more people as elected by you to handle your finances/property and personal care when you are still alive but unable to make decisions. This essentially gives these people the ability to *be* you. You need to seriously consider who you choose to give this power to.

Due to the weight of this, I don't recommend the home-kit versions. A good lawyer will give you valuable advice that will more than cover the cost in the long run. This document becomes null and void once you die, because once you die, your will takes over.

Executor

An executor is the person who makes sure your affairs are wrapped up after you die. You must name this person in your will. This individual will sell your assets to pay your debts and taxes, and make sure that what's left goes to whomever you wanted it to. This might be challenging if you didn't leave a list—which, again, is what your will is.

One thing to note on this as well is that it is possible to name an estate-planning company as your executor—a Corporate executor. Being an executor is a very challenging role, and most of the time we name our spouse or child. Yet at the time of your death, that person will be busy trying to grieve. Would you rather that individual have the time to spend with family in remembrance, or spend it toiling over your credit and investment statements trying to make sure that all the t's are crossed and the i's are dotted in your final income-tax return and then distributing your assets to the family?

Before you name your spouse or your child as the executor, find someone you know who has been an executor of an estate and ask that person how easy it was. I am willing to bet that you will find very few who say, "Oh, that wasn't a big deal, pretty simple actually." You are more apt to hear about the hours upon hours required to get everything sorted out and how they still weren't 100%

positive that they got everything at the end, but since they haven't heard anything about it since, they must've done okay. Most corporate executors do require that the estate be of a certain value, so this might not be an option for everyone, but you should know the possibility exists.

Until you have been an executor of an estate, you really can't understand how challenging it is to wrap up another person's affairs. Why would you want to make it more difficult for your loved ones by not writing everything down in the form of a will?

I always use my situation as an example. If I die tomorrow, my ex-wife would celebrate … I mean, be traumatized. If I didn't have my endgame laid out in detail, she would have to take care of paying my debts, distributing my assets, paying my income taxes and a whole host of other things—while still finding time to party in the Florida Keys … or mourn. While all of this was going on, she'd also need to console the children and help them heal. Since I was always an @$$, I didn't write anything down, so now she'd have to decide what to do with each piece of everything I own. Sounds like a big job. Wouldn't it be nice of me to leave her a list in the form of a will?

Before that even happens, she would have to sort out the details of the wake and the funeral: What kind of flowers? Where is the service to be held? How many people should be invited? Who should be invited? Do I want to be buried or cremated? In the middle of all this, when is she supposed to find time to grieve? So many of us just ignore this, but why? Do we just want one last opportunity to be a pain in the neck to our family?

Insurance agents will tell you to take out additional insurance to pay for your funeral expenses. Perhaps there is some validity to this, but your family still has to do all the planning after you are gone. Perhaps prepaying the funeral directly and planning this while you are still around also has some merit.

Estate tax

Estate tax is an area that many people would like to circumvent by limiting or eliminating the assets in the estate that are subject to probate. Probate is, in simplest terms, a legal process that certifies that your will is your final will and it is valid to be acted upon by your executors.

Avoiding probate is not always wise, but it may work in some situations. There are a few ways to do this. You can make bank accounts, investment accounts and property joint with the people who you want to have it when you're gone. The challenge with doing this is that they can also access it while you are alive. Spendthrift relatives might empty your account before you're gone, affecting your ability to live the lifestyle you planned for. There are many other reasons why sidestepping probate might not be a great strategy, but there are ways to do it, if it's right for you.

I have seen many clients over the years who are widows/widowers with only one child. Their desire is to leave everything to that child, so they make all of their assets joint (with right of survivorship) with that child. If you have many beneficiaries, one option is to give them some of their "inheritance" while you are still alive. Gifts of property or money can be passed to the next generation, and you can spend part of your later years watching your heirs enjoy their inheritance while you are still here.

There are numerous other estate-planning techniques, especially for larger estates. For these types of strategies, it is strongly recommended that you consult an Estate Planner so that your plan is tailored to your unique situation.

10 Protection

Not everything that can be counted counts,
and not everything that counts can be counted.
—*Albert Einstein*
Physicist

Every day, we take risks. When we wake up and get in the car to go to work, we risk the road. Everything we do or don't do has some form of risk related to it. We have the choice to accept these risks or protect ourselves against them. Many people seem to feel that the only types of protection they should get are the ones they are forced to have, like car insurance and fire insurance on your house when you have a mortgage.

In its simplest form, insurance is an agreement between you and an insurance company that says if something happens to you, the company will pay you (or someone else) some money. This could be a critical illness, disability or loss of life, to name a few different kinds of insurance. There are many others.

An individual policy is an agreement between you and the insurance company. Typically, when you are completing the application, you are asked to answer some questions and donate some bodily fluids. This is done so that the company can determine how healthy you are at that time. Depending on the type and amount of insurance you are asking for, your answers to the questions and the results of a battery of tests performed on those bodily fluids will tell the company whether or not to give you the coverage you want—and how much it will cost you to have the agreement put in place. This is called *underwriting*.

A group policy is an agreement between an insurance company and a group that you belong to. Usually group coverage is less expensive, as the insurance company is getting a large number of people to sign up and probably only has to pay out to a few. The company usually only asks a couple of simple questions and rarely does any tests on your health. The underwriting of a group policy is done, usually, only when a claim is filed.

One advantage of individual coverage is that you know the policy will pay out. One advantage of a group policy is that you would likely have a lower cost—and as long as you don't have any sort of pre-existing condition, the policy will likely be honoured.

Credit protection is a form of group insurance that everyone who has credit with a bank pays into if they accept the coverage. When something happens to you like job loss, critical illness, disability or death, the insurance pays the balance off or makes monthly payments for you for a period of time. This helps protect your credit rating and ensures that your current lifestyle and your estate are preserved.

In this form, credit protection serves to eliminate your debt while your personal insurance (term, whole life and universal policies) are in place to act as income replacement or transfer of wealth to the next generation. I suggest that both personal insurance and credit insurance have their place in a properly planned protection strategy.

When most people think of insurance, they think of life insurance. Oftentimes people will tell me, "At that point I'm dead. Who cares?"

My point is—what if you live?

Suffering from a disability or critical illness not only reduces your income because you aren't able to work, in many cases it also increases your expenses due to medical costs. There are insurance vehicles that provide disability or critical illness protection. In Canada, we often take this for granted due to the health care system that we enjoy. The longer I am in the financial industry, the more often I see situations where people have to go abroad for medical procedures, or they discover too late that their work coverage really isn't as good as they thought it was.

I had a couple who I assisted with a refinance of their mortgage. They said no to the credit protection. Two months later, the husband suffered an injury at work, needed immediate surgery and was off work for eight months recovering. A couple weeks before, his wife had changed jobs. During his recovery period, money was tight, because he was only receiving a partial income from employment insurance. Before his wife was done with her probationary period at the new job, she was diagnosed with cancer, and the chemotherapy had her off work as well. They had to sell their house because it was no longer affordable. Disability coverage on the mortgage would have made the mortgage and property-tax payments for them when he got injured. Critical-illness coverage would have paid the mortgage balance in full when she was diagnosed with cancer. All of this happened between February and August.

I don't think I need to tell you how badly this can affect the financial plan of a couple in their late forties.

We can't predict the future, but we can certainly prepare for it. The right kind of protection strategy can provide you with time and options. You don't want to be in a position with no time and no options. Make sure that you have professional guidance in preparing your protection strategy.

11 Taxes

You must pay taxes. But there is no law that says you gotta leave a tip.

—*Morgan Stanley*
(Advertisement)

WHEN WE ARE alive—and when we die—one of the largest expenses that we have is tax. In Canada, our income-tax system is tiered. This means that when you earn more, you are only taxed a higher rate on the amount of money that falls into that tax bracket. These brackets and the amount of taxes charged change periodically. This applies to both federal and provincial/territorial taxes.

If you want to understand more about how much tax you are paying and get access to a number of great tools for calculating the benefit of RRSP contributions on your income as well as a variety of other things, Google "Ernst and Young tax calculator" or look at the websites I have included at the end of this book. I have used their tools with many of my clients over the years.

Personally, I am okay with the idea of paying taxes. They pay for our fire departments, police, road maintenance—the things that make our country a safe and enjoyable place to live. However, I don't believe in paying more taxes than I am required to.

Legal ways to pay less tax

From a financial-planning perspective, there are a few things you can do to pay less tax ... legally. You can earn less, for one thing. Perhaps that is not the best financial planning strategy, however. You can move to another province that has lower provincial taxes, but this also may not be a feasible option for many people. Let's have a look at a few strategies that can be used to reduce your tax burden legally and put another ace up your sleeve.

Minimizing taxes on non-registered investments

As we have discussed already, taxes have a big impact on our lifestyle. Let's say that over the course of your life, you save and invest, and over the years you accumulate $100,000, which you decide should be placed in a guaranteed investment certificate (GIC). Let's say that you get an interest rate of 4%. This 4% would give you an income of $4,000 per year, which you would have to pay taxes on. Let's say you are in a 40% tax bracket, so you have to pay $1,600 in taxes; now you have $2,400 in income, which is really a 2.4% return. So here, 40% of you income is going to taxes.

What we could do is consider a different type of investment, but there is more volatility to it. If you were to invest $100,000 in a fund that was selling at $10 per unit, you would purchase 10,000 units. One year later, there is a 4% increase, and your units are now worth $10.40 each, which means you now have $104,000. In order to have that $4,000 income, you would sell 385 units at $10.40. In this scenario, you pay taxes on the gain due to the nature of the investment. The tax implication is as follows:

$$385 \times \$0.40 = \$154 \text{ gross capital gain,}$$
$$\text{which is only 50\% taxable, so}$$
$$\$154 \div 2 = \$77$$

You are still in a 40% tax bracket, so you would have to pay 40% of $77 = $31 (rounded) in taxes. The difference in your pocket looks like this:

- You have $4,000 of income from a GIC, you pay $1,600 in taxes and you have a net income of $2,400.
- You have $4,000 of capital gains income, you pay $31 in taxes and you have a net income of $3,969.

Which would you prefer?

RRSP versus Leverage

People in a higher net-worth position may want to consider the advantages of using leverage (borrowing to invest) as part of their investment plan and as a tool to grow your net worth. Please review leverage disclosures that you can get from your advisor or online and consult your tax advisor and financial advisor to determine if these strategies may be appropriate for you.

One of the reasons I feel leverage makes sense for some people is this: Let's say you are going to invest $1,000 a month or $12,000 a year in an RRSP. Over 15 years at 6% your investment grows to $288,308. An alternative to that is to take out a $200,000 line of credit at a 6% interest rate. With interest-only payments, that would make your payments the same $1,000 per month. This money would be invested in a non-registered investment. That $200,000 grows at 6% over 15 years, and the value at the end is $479,312.

During the 15 years, the interest that you pay on the line of credit is tax deductible like an RRSP, and if you cash out part of the investment to pay off the line of credit at the end of the 15 years, you end up with basically the same amount of money. So you've got the same amount of money out of pocket each year, and the money is invested over the same period of time, earning the same return. The difference is that when you take money out of an RRSP, it is 100% taxable. If the return on your non-registered investment is capital gains, it's only 50% taxable.

RRSP Meltdown

There are many Canadians with very large RRSP portfolios. As we've discussed, this is a great benefit when you are working and getting refunds on your taxes, but later in life when you need to withdraw from your RRSPs, it can carry a heavy tax burden, and it may also impact certain government income supplements like Old Age Security. The RRSP Meltdown strategy allows you to withdraw all of your money, over time, from your RRSP tax-free.

Let's say you have a $200,000 RRSP and you want to take it out over 10 years between ages 55 and 65. This means you would withdraw $20,000 each year, and you would have an additional $20,000 in income to claim on your taxes each year during that period. If you take out a $400,000 line of credit at a 5% rate with interest-only payments, the annual payments of $20,000 are tax deductible. Now you have an additional $20,000 income, but you also have an additional $20,000 tax deduction. One nullifies the other from a tax perspective. But now you have a $400,000 non-registered investment, and by the time you are 65, all of your RRSPs have been taken out. You don't have to worry about being forced to take additional income at any point in the future like you would if the money was left as an RRSP.

One of the other considerations regarding paying less tax is whether you can implement an income-splitting strategy at the same time. Spousal RRSPs are great for this, as discussed in Chapter 7.

Having a tax-planned will and intelligently managing the estate tax are options for reducing the taxes you pay on your way out the door. Due to the nature of these types of strategies, their complexities and how geared they are to your unique situation, I will not delve into them further. I will simply suggest you do your own investigation and consult the team of advisors you are building—your accountant, lawyer and financial planner—to work with you to devise the best plan.

At the end of the day, it really is not about how much you earn. It's about how much you keep. Proper tax planning with your assets and investments will make a significant difference in your net worth and your lifestyle.

12 Owning versus Renting

Twenty years from now you will be more disappointed by the things that you didn't do than by those that you did do.

—*Mark Twain*
Author

THAT IS PROBABLY the quote that rings most true for me, personally. Twenty years ago, I had the opportunity to buy a property. Twenty years ago, I did not buy that property. There were all kinds of reasons why I didn't—one of them being that everyone I spoke to was saying, "Don't buy, the housing bubble is going to burst." The one person who was saying *buy* was my spouse. Twenty years later, I still feel the pangs of not buying that property. Twenty years later, people are still saying the housing bubble is going to burst. Twenty years later, she is still saying she was right, and she was.

Of the clients I have seen over the years, the most financially stable are those who bought their first home early in life. The purchase of a home is one of the largest decisions most people will make in their lives. It is critical that you have the appropriate information to make an informed decision. There are very few reasons, in my opinion, why a person would not purchase a home—for example, if you have very low rent, if you have employer-paid accommodations, if you relocate every year or two, or if you have a very short life expectancy. The advantage of renting is that you don't have to worry about the time and expense involved with upkeep and repair, which may be suited to someone who is regularly abroad.

To start on your home-ownership venture, the first item to take into consideration is the down payment and closing costs.

Down payment and closing costs

At this time, in Canada, home buyers can purchase their primary residence with as little as 5% of the purchase price, up to certain home prices. However, you also need to ensure that you have an additional 1.5% of the purchase price in savings to cover closing costs. These are things like legal fees, land-transfer taxes, moving expenses and relocation cost of your utilities. It's also not a bad idea to have some extra savings for things like the new couch, TV, washer and dryer that inevitably you will find you want to buy as soon as or shortly after you move in. Having extra money is never a bad thing.

If you have a goal to purchase a $300,000 home, you will need at least $15,000 for the down payment and $4,500 for closing costs. If we go back to the goal triangle I talked about in Chapter 2, we can put ($15,000 + $4,500 =) $19,500 as the goal in the middle. Time will be the length of time until you want to buy the home, and $ is the amount you are dedicated to setting aside weekly, biweekly or monthly to save that $19,500.

Since the process of saving for a down payment usually involves a short time frame, I would recommend that the return on investment for this particular goal be maintained as whatever the savings account interest rate is at the time you are saving. It is advisable that goals within such a short time frame not be aggressively invested, as a downturn in the market can add significant additional time to achieving the goal, which will only hinder your longer-term desires.

Title insurance

If you did purchase this $300,000 home using only 5% down, the mortgage company you borrowed the $285,000 from would have to apply for title insurance from either the Canadian Mortgage and Housing Corporation (CMHC) or Genworth Financial. Federal regulation in Canada stipulates that anyone purchasing a primary residence with less than a 20% down payment must acquire title insurance. This insurance, in its simplest form, guarantees that if the owner of the home doesn't pay the mortgage and the bank is forced to take the house and sell it (foreclose), and the bank is unable to sell it for at least the balance on the mortgage loan, the insurance kicks in and covers the difference between the sale price of the home and the balance of the loan. This ensures that the bank doesn't lose money in the deal. This is one of the reasons why Canadian banks are considered the most stable banks in the world.

The cost of this insurance will vary depending on the percentage of down payment that you have. For example, if you have 5% as your down payment and you choose a traditional 25-year amortization on your mortgage, the fee is 2.75% of the purchase price. This fee is added to the balance of the mortgage, and the lender pays CMHC directly. By adding the title insurance to your mortgage, you increase the long-term cost of the mortgage, so you may want to consider having the cost of the fee saved and, shortly after the mortgage is funded, apply a principle payment to the mortgage for the amount of the CMHC fee, effectively eliminating the cost of paying interest on the fee.

Being able to purchase a home with a small down payment provides you with the flexibility to purchase your first home sooner. As we know, life happens. The longer you are saving for the down payment, the more likely that something else will come up that causes you to dip into your savings and continue to push the goal further into the future.

Get your rent back

Many people feel that their home is the greatest asset they will ever acquire. For many, this may be true. I would also suggest that for these same people, it is likely the greatest debt they will ever have. The way I look at it in most situations is, you can rent from landlords and increase their net worth, or you can rent from the bank and increase your own net worth. The cool thing about renting from the bank is that, when you move, you get your rent back.

When you sell your house—even if you sell it for the same price that you bought it for—you are effectively being returned your monthly payments because your mortgage balance has come down. In this case, you might be out-of-pocket the cost of the interest on the mortgage loan, but you are getting most of your money back. Have you ever rented and moved and tried to ask your landlord for your any of your rent back?

Mortgage strategies

A lot of people seem to have an affinity for a fixed five-year rate on a mortgage. I am not sure why, especially since CMHC research indicates that the average Canadian refinances to renovate, consolidates or sells a home on average every 34 months, ending up with blended interest rates or penalties or having to deal with the same bank to avoid these things. But that's a rant for another day.

Let's look at a situation where you have a $350,000 mortgage and you are looking at a one-year fixed rate of 3.29% or a five-year fixed rate of 4.49%. Yes, the fixed five-year rate means you have a long time before you have to worry about rates going up. But let's look at an alternative option.

If you have a $350,000 mortgage with a one-year rate of 3.29% and you amortize it over 25 years, your principal and interest payments are $1,713.01 per month. If you took the five-year rate of 4.49% your payment would be $1943.42.

If you take the five year rate with the $1943.42 monthly payment, assuming nothing changes, in 25 years you are mortgage-free. Hooray!

Here is what I would do. Take the one-year mortgage and the $230.41 per month that you are not paying to your mortgage set that up as a regular investment plan into your RRSP. That gives you an annual RRSP contribution of $2,764.92—and if you are in a 40% tax bracket, that equals a $1,100 tax refund. This tax refund I apply to my mortgage as a principal payment annually. It pays my mortgage off in 23 years instead of 25 and saves me more than $14,000 in interest. If I switch to biweekly payments of $1713 ÷ 2 = $854, then my mortgage will be paid off in about 21 years, and I save more than $32,000 in interest. Meanwhile, my RRSP of $230.41 per month is growing at 4%, and in 25 years it grows to more than $117,000.

In the first scenario, you pay $1943.42 per month over 25 years, which is $582,975 to pay off a $350,000 mortgage. In the second scenario, you pay $854 biweekly over 21 years, which is $466,284 to pay off the $350,000 mortgage and end up with an RRSP of $117,000. Both ways, you are spending $1943.42 per month, but the end result is phenomenally different depending on which strategy you use.

The best way to set up something like this is to automate it. I will continue with the above example. Every two weeks, when I get paid, I am going to have $975 transferred from my transactional account to my savings account. My mortgage of $854 comes out of that savings account. I also set up my RRSP of $230 per month, which I changed to biweekly too just to make it easier, so $115 every two weeks goes from my savings account to my RRSP. Most banks allow RRSP investments and mortgage payments to be debited from a savings account without fees. Check with your financial institution to verify this; otherwise, a low-transaction account will do fine, and your net benefit will substantially outweigh the few dollars for a second transactional account. I recommend setting it up this way because most people see the automatic transfer coming out of their account and just think of it like it's their mortgage payment and are more likely to keep the plan in place.

One other strategy with mortgages is, when you do need to refinance, rather than taking a 10- or 15-year amortization, extend it out to 25 years and repeat the steps above, but the amount you have set up to transfer automatically to your savings account is the mortgage payment for the 10- or 15-year amortization that you want. Using the same $350,000 mortgage above, with a 15-year amortization you would be putting about $1,350 every two weeks into the account, and the same $854 goes to pay the mortgage. Now you have $500 every two weeks going into your RRSP, plus a $5,200 tax refund which gets applied to the principal of the mortgage. The mortgage gets paid off in 15 years, and with a 4% return in 15 years, you now have a $245,000 RRSP.

The really big advantage here is that if life happens and your income goes down, you can stop the RRSP contribution, and the mortgage payment is $854, which is a lot more affordable than the $1,350 you were planning on when you were asking for a 15-year amortization. These strategies give you a substantial amount of benefit in both good times and bad.

13 A Final Word

It is not the strongest of the species that survive, nor the most intelligent, but the one most responsive to change.

—*Charles Darwin*

Naturalist

Generally speaking, people are motivated by fear and/or greed. If you properly plan your financial future in small segments of one to five years, and adjust or tweak the plan periodically as life changes, you can easily discover that these two factors are virtually eliminated, and you can make decisions based on your goals, logic and common sense. I might suggest that it is likely that you will make better decisions under these circumstances than if you are being driven by fear or greed, thinking, *I have to do this*, or *I can't afford to do that*.

Once you have a clear ability to make a plan by reducing the noise that daily life throws at you, then you can effectively develop and design your path forward through the financial mazes that you will explore during your progress from dependency to luxuries or wherever in between you feel you want to wedge your standard of living. If you do things the way you were brought up to do them, that may no longer be the right way to do them. Today you will start with the new way to do them.

I don't want you to make the same mistakes most Canadians make, because by failing to plan you are planning to fail. Today is your day for change. Today is your day to take control of your financial future. You haven't come this far to learn all these things to do nothing with it.

Now that you understand the game you are playing, the basic rules of the game and what cards are in the deck, it's time to play the game. Start by listing all of your monthly expenses and their associated costs—that is, create a budget.

Do you have money left at the end of the month? Great, see B. Do you have month left at the end of the money? That's okay, see A:

A. You need to stop the leaks that may inevitably cause your ship to sink. Go over your budget and see what you can reduce or eliminate from your expenses. The first thing you need to do is to make sure that you have more money coming in than going out. It may help to talk to someone.
B. Leftover income should work for you, not be idle. It is better for you to have a future vacation than to allow your money to take a current one.

Now, list three financial goals that you have, and *let's make them happen!*

1.
2.
3.

With each goal, determine the dollar value and importance of each to your lifestyle. Take a look at the tools and strategies provided in this book and then find an advisor you feel comfortable with and who is willing to help you reach these goals.

Depending on the dollar value and time involved, it may be as simple as "paying yourself first." Sometimes it may mean a trade-off, focussing on what you want *most* instead of what you want *right now*. You may have to trade the latte today for something shinier tomorrow.

Remember this is (supposed to be) the best part. You have goals. You have a plan. If you are tracking, measuring, monitoring and periodically tweaking your plan and have an appropriate protection plan in place, you are almost guaranteed to be successful. Set up your accounts with a label indicating what the money is for. That way, when you use your online banking, you are reminded of the progress you are making towards that goal.

When you are ready to implement more complex strategies, you will benefit from the ideas in this book combined with the help of an advisor. Pick up the phone and make that appointment. The strategies and concepts that I have shared with you throughout this adventure have been developed over years. These methods have made remarkable differences in people's lives. I strongly recommend adding as many of these cards as you can to your deck. They will need to be adjusted to suit your needs, and a qualified financial professional can help you do that—or you can also DIY if you feel that you need to manage things.

I do recommend, as with everything, that you use caution when deciding how much and what aspects of your finances you want to self-direct, because if anything happens to you, who is managing your money while you are incapacitated or during the months or years that it takes to execute your estate? There is a lot of value to having a team of professionals provide you with support and expertise.

Now I need you to be forewarned: Not all of these strategies are appropriate for everyone. There are benefits and risks associated with each of these ideas. Conversely, there are benefits and risks to doing nothing. You need to decide which path you want to take, and if the numbers presented use dollar values that are too high or too low for you, simply move the decimal place; the concept and math remains the same.

Some of the methods and examples used in this book involve the concept of borrowing to invest. The Mutual Fund Dealers Association of Canada (MFDA) provides leverage risk disclosures, and it is a requirement for all MFDA licensed representatives to provide investors with this information. It is a good idea to be familiar with the risks associated with leverage. I recommend that you go to the MFDA website to read the disclosures on this topic.

In my opinion, personal finance is one of *the* most important things in your life. The majority of Canadians, historically, haven't paid much attention to it. Those who do become successful financially. Those who don't, don't.

If you want to be successful in your financial life, you need to do the following:

1. Understand your goals.
2. Understand your comfort zone.
3. Set reasonable targets and tweak them periodically.
4. Plan long-term.
5. Have a safety net.

Do these five things and implement even one of the strategies from this book, and I guarantee that you will be richer tomorrow than you are today. Stick to one of these strategies long-term, and I guarantee that you will recoup more than a thousand times what you spent on this book.

Appendix 1

KNOW YOUR FINANCIAL SELF

The point of these questionnaires is for introspection to your individual circumstance. They are also meant to give you a starting point to understand where you are financially. The category descriptions should help you understand what you may need to do in order to move your wedge up the Standard of Living Hill and secure it in the place you want it to be.

Where's My Wedge?

1. To better understand finance, I use the following resources:
   A. None
   B. The lady at the bank takes care of it.
   C. I do some research and talk with an advisor.
   D. I use all available resources and monitor my finances regularly.

2. Since last year, my financial situation has:
   A. Improved.
   B. Stayed the same.
   C. Gotten worse.
   D. I'm not sure; I haven't been keeping track.

3. In the coming year, I believe my finances will:
   A. Be about the same.
   B. Probably be worse.
   C. Likely be better.
   D. It's impossible for me to be sure either way.

4. Do you have more assets and income than debts and liabilities?
   A. Yes
   B. No
   C. I don't know

5. How big is your cushion or emergency fund?
   A. 4–6 months
   B. 3 months
   C. 2 months
   D. I don't have one.

6. How financially educated do you feel?
    A. Global expert
    B. Formally educated with experience
    C. Some basics and a budget
    D. Getting an idea
    E. Just out of the nest/not really

7. Do you plan and use a budget?
    A. Yes to both
    B. I budget
    C. No

8. What kind of decision-maker are you?
    A. I make informed decisions.
    B. Others make decisions for me.
    C. I make decisions with some information.
    D. I make spontaneous decisions.

9. Do you have a positive or negative cash flow?
    A. Positive
    B. Negative
    C. I don't know

10. How often do you review your financial situation?
    A. Regularly—more than once a year
    B. Once a year
    C. Never

11. Which of these describes your net worth?
    A. Positive
    B. Negative
    C. I don't know

12. Are you aware of where your money goes?
    A. All of it—essentially every cent
    B. The majority of it
    C. Only my bills
    D. Don't track any of it

13. Do you know your credit score?
    A. Yes
    B. No
    C. I have a general idea

14. Do you know what your credit bureau report looks like?
    A. Yes
    B. No

15. What is your total debt service ratio?
    A. I have no idea.
    B. Less than 30%
    C. Less than 40%
    D. Too high

16. How well do you understand the financial products you use, such as your bank account or loans?
    A. I've read the fine print.
    B. I use them and know the basics.
    C. I just utilize them.
    D. I don't have any financial products.

17. Are you cost savvy? Do you know and understand what fees and rates you pay for your financial products as well as any early payout penalties?
    A. No idea
    B. I am aware of monthly fees and rates.
    C. I know about my fees, rates, penalties and any additional possible costs.

18. Do you have a good idea of the financial products available in the market and their possible uses and benefits?
    A. Not really
    B. I am aware of some options for my finances.
    C. I am aware and I understand many of the tools and products available to help with the development of my financial plan.
    D. I understand and can explain to others the features and benefits of most financial products available in the market.

19. How is your debt structured?
    A. Don't have any.
    B. Long-term debts are at low rates, credit cards are paid in full monthly.
    C. Credit cards usually have a balance, but I make my payments on time.
    D. It is messy, or I really don't know.

20. What does your savings strategy look like?
    A. Still paycheque to paycheque.
    B. I save a little here and there when I can.
    C. I have saved quite a bit, and I think I'm doing all right.
    D. Automatic savings in place, and my money is invested in a way I am comfortable with.

21. Which umbrellas are you using for your investment plan?
    A. Umbrella? Is it raining?
    B. Only RRSPs
    C. Only TFSA
    D. RRSPs and TFSAs
    E. RRSPs, TFSAs, non-registered, and it's all automatic

22. You are currently using which of the following investment vehicles?
    A. Haven't started investing yet.
    B. I have some in a savings account.
    C. GICs are my focus.
    D. I have some mutual funds.
    E. I have a wide range of investments, including all of the above, real estate and other investment vehicles as well.

23. What were the results of your investment profile?
    A. Conservative
    B. Moderate
    C. Balanced
    D. Aggressive
    E. Growth

24. Which strategy are you using for your RRSPs?
    A. None
    B. I have a company pension, so I'm not worried about RRSPs.
    C. Automatic investment plan and putting in as much as makes sense based on my plan
    D. My maximum allowable limit

25. Have you registered for access to your Service Canada account?
    A. No
    B. Yes, but I really don't use the site.
    C. Yes, and I check it periodically as part of my plan.

26. Are you on track to achieve financial freedom?
    A. Haven't started yet
    B. Fell off the track, and I am trying to get back on the right path.
    C. I know what it looks like, and I am working on it.
    D. I am already there. Work is optional for me.

27. Are you currently satisfied with your plan to achieve your top three most important financial goals?
    A. Quite frankly, I don't have a plan.
    B. I know my goals, and I sort of have a plan.
    C. My goals and plans are clear; I just need to work on them a bit more.
    D. The plan is working well, and I review it regularly.

28. My Tax-Free Savings (Investment) Plan is:
    A. Not set up
    B. I have one, but I'm not sure what it's doing.
    C. It is invested and diversified in line with my plan.

29. As far as estate planning goes:
    A. I don't have a plan.
    B. I have a will, but it might not be up to date.
    C. I have a will and power of attorney.
    D. I have all my documents in place. They are current, and I have a detailed plan of how everything should be managed for me.

30. Which protection and risk-management strategies do you use?
    A. None. I don't believe in insurance.
    B. I have some coverage at work, but I can't give you the specifics of what it does for me.
    C. I have income replacement strategies as part of my plan to protect me and my family.
    D. I have B and C, plus I have debt protection coverage in place as well.

31. As far as income taxes go:
    A. I have no idea how much I pay.
    B. I generally understand my taxes, but I don't know what to do to change that.
    C. I know I am paying more than I need to, so I try to reduce it as much as I can each year.
    D. I have maximized my tax-reduction options for both the present and the future.

32. What are your current living arrangements?
    A. I don't pay rent.
    B. I live on my own and pay rent.
    C. I own my own property with a mortgage.
    D. I have my own property, and I have rental income.

33. Are you willing to sacrifice some of what you want today in order secure a better tomorrow?
    A. No, I prefer to live in the now.
    B. Maybe, but I need to be able to see concrete results and will need reminders of the benefits.
    C. Yes, absolutely. I understand the value of enjoying today while planning for tomorrow.
    D. I've been doing that for a while, and because of that, I am in good shape going forward. Even so, I am still reviewing and tweaking my plan regularly.

34. How motivated are you to achieve your financial goals?
    A. Not overly
    B. I want to do something, but the situation seems pretty bleak right now, and I am not sure where to begin.
    C. I am very motivated to realize my dreams.
    D. I am motivated and well on my way, with a comprehensive plan that I review regularly.

35. If you could change only one thing about your financial situation, what would it be?
    A. Cash flow
    B. The amount of interest I pay on my debt
    C. The amount of taxes I pay
    D. Increase my assets
    E. I just need a plan.

Scoring Grid for Where's My Wedge?

Question	Point value				
	A	B	C	D	E
1	0	2	4	6	
2	6	4	2	0	
3	4	2	6	0	
4	4	2	0		
5	6	4	2	0	
6	8	6	4	2	0
7	4	2	0		
8	6	2	4	0	
9	4	2	0		
10	4	2	0		
11	4	2	0		
12	6	4	2	0	
13	4	0	2		
14	4	0			
15	0	6	4	2	
16	6	4	0	2	
17	0	2	4		
18	0	2	4	6	
19	6	4	2	0	
20	0	2	4	6	
21	0	2	2	4	6
22	0	2	2	4	6
23	4	4	4	4	4
24	0	2	4	6	
25	0	2	4		
26	0	2	4	6	
27	0	2	4	6	
28	0	2	4		
29	0	2	4	6	
30	0	2	4	6	
31	0	2	4	6	
32	0	2	4	6	
33	0	2	4	6	
34	0	2	4	6	
35	4	2	6	8	0

What your scores mean

The following categories are not hard and fast rules. As with most of the rest of this book, the results from this are generalizations that will apply to the majority of the population. You may find that you have one foot in one category and one foot in another.

Score 0–49
Dependent

There are two different kinds of dependent. The first one I partially explained earlier, and here is a recap. You are probably young, likely still living with family and possibly still in school. Being dependent at this point in life is to be expected.

If that is who you are, and you are reading this, the world is your oyster, and your financial future is phenomenally bright, because time is on your side. Right now, you should open two bank accounts (if you don't have them already) and start saving money. Have another look at Chapters 5 and 6 for some tips on automatic saving and the power of investing early. Birthday money, holiday money, part-time-job money—wherever it comes from, save some.

Make one account your spending account and one your saving account. You have no real expenses at this point, so put half your money in your spending account and the other half in your saving account. Once you are 18 or 19, have a chat with an advisor, develop a bigger plan and implement the strategies in this book. Good job! You're off to a great beginning because you are starting early.

The second class of dependent tends to not necessarily be dependent on others but rather reliant on credit or in the habit of simply doing without. If you are in this category, you are an adult, likely employed and possibly married with more than one child. You are living in the situation where the money comes in and is spent in a few days, and you have a week to ten days before the next paycheque.

The first thing you need to do is to *want* to change your financial situation. Actually, you probably need to make it your number-one priority. Using the tools in Chapter 2, make a budget, follow it, understand where all your money is going and then look for ways to decrease your expenses by spending differently.

Also give consideration to creative ways to increase your income. Maybe you should look at doing what John and Donna did in Chapter 2. Most people in this category don't have assets to borrow against, and they often have a few blemishes on their credit history, so consolidation may not be an option. If it is, use it. Also, go back to Chapter 3, look for things that may be bringing your credit rating down and look for ways to improve your credit.

Your road to financial success will require a lot of intestinal fortitude. Stay strong, stay focused and stay 100% committed. In time, you'll see the light at the end of the tunnel and the warmth of the sun on your face when you get to the end. This will be so rewarding and such a relief that you will feel a great sense of pride in this accomplishment.

Score 50–111
Independent

Individuals in this category usually have some interest in and understanding of finance. However, they are usually still living paycheque to paycheque. Cash flow is close to break even or even a small positive, but many of these people run out of money the night before payday.

Independents also tend to offset cash flow shortages by overusing overdraft protection, credit cards or lines of credit—and as a result may be carrying a lot of debt. Be wary of this.

Independents typically don't have much in the way of savings, and their net worth is likely low. You may use a budget but likely don't watch your spending closely, and you still probably make spontaneous decisions periodically. You possibly don't have a great understanding of financial products or your credit.

Most people in this category don't have a written financial plan, nor do they have a scheduled review of their finances. You may be good at paying your debts on time, and you may even pay off your credit cards regularly, but you need to make sure that you are not spending more than you make. Go back to Chapter 2 and use the budget worksheet and cash flow statement to see if you can open up a bit of cash flow. A consolidation might be a good idea.

Review Chapters 2, 4 and 5 to make sure you understand what you're doing and what other options there may be.

I know the savings aspect is hard, but you *have* to do it too. If you are waiting to start saving until you have paid off your debt, it will likely be a long time before you begin, because life happens. Build your safety net and then start acquiring assets that appreciate in value. That is your key to moving to the next level.

Score 112–149
Quality

A quality lifestyle typically means that you do some of your own research and planning and probably consult a financial professional. Your financial situation is likely improving year over year and will probably continue to do so in the future. Your net worth and cash flow are positive, and your debt is probably well organized for the most part. You most likely have a decent cushion/emergency fund to help protect you against the unexpected.

People enjoying a quality lifestyle typically use a budget and make informed decisions regarding their finances so that they know where their money is going. They tend to review their situation periodically—but perhaps only at mortgage renewal time or if their GICs are up for renewal. You probably know your credit is decent, but you likely don't monitor your credit report. Part of a quality lifestyle means that you are roughly aware of the rates you are paying and what costs and fees are associated with your finances because you have a fairly good understanding of financial products and services. The biggest opportunity here is to make sure the money you work so hard for is working hard for you.

Have another look at Chapters 6, 7 and 8 for ideas to help make sure your money is invested, and Chapter 11 for recommendations on when you need to borrow money.

Make sure that you take advantage of the equity in your property. If you want to move to the next level, you will want to focus on growing your net worth and exploring options for passive income. It would also be a good idea to review your plan more frequently.

Score 150–179
Comforts

At the comforts stage, life feels pretty good. You may not spend a lot of time thinking about your finances because you always have money. There is always sufficient money for regular lifestyle expenses, with enough left over that you also don't really have to think about those little splurges or "just because" gifts. You definitely have some great equity and a nice nest egg of savings and investments.

Your understanding of financial products, fees and rates is above average, and you are probably rate-sensitive. This means you like to shop around for financing options for big-ticket items. You probably have, or have had, rental properties. You likely deal with more than one financial institution because you feel that ensures that you don't have all your eggs in one basket.

Many people at the comforts stage run into difficulties if there is a disability or long-term illness. Have another look at Chapter 10 on protection, and make sure that your protection strategy is well developed so that you don't have to liquidate assets if you get hit with a curveball. If you want to move to the next level, not only do you need to have that downside protection, but you also need to make certain that your assets are staying ahead of inflation. To do this, make sure that your portfolio of assets is diversified.

Finally, you are possibly paying more taxes than you need to. Review Chapters 9 and 11 on estate and tax planning and then gather your team of professionals to help reduce your tax burden and get your will and power of attorney in place. Then make sure that these are updated regularly along with the rest of your plan.

Score 180 +
Luxuries

Congratulations! You're at the top of the Hill. This is where the majority of the population wants to be. This is the type of lifestyle we typically see portrayed in the media and flashing across screens for visual entertainment. For many, it's the "I wish" category.

It probably took a lot of focused dedication to get to where you're at. You've been smart with your money and likely have taken a few calculated risks. You have a team of trusted advisors for many different aspects of your finances, and they support you in making informed financial choices. Your cash flow is maximized, and your net worth is substantial. If you aren't at the point where work is optional, you are close and on track to getting there soon.

You regularly check up on your finances, and you have a solid grasp of what you are doing and why. You know the kinds of fees and costs associated with your plan, but you don't always look for the cheapest option because you know that sometimes you get what you pay for and that you have to spend money to make money. If you have any debt, it is probably tax deductible. You have a wide range of financial products and a diverse array of investment and income-generation options.

Your area of opportunity is likely from Chapter 9 on estate planning and transfer of wealth to the next generation. Make sure that your will and power of attorney are current. Consider the possibility of having an estate protection strategy. You may also find value in the RRSP Meltdown strategy in Chapter 11. As long as those things are in place and you keep doing what you are doing, you have probably secured your position in this stage for the long term. How much more you grow your wealth is entirely up to your desire.

Investor Profile Questionnaire

Question	Point value

Personal and Financial Situation

1. Age:
   - Under 30 — 8
   - 31–40 — 6
   - 41–50 — 4
   - 51–65 — 2
   - Over 65 — 0

2. Income:
   - Under $40,000 — 0
   - $40,001–$75,000 — 4
   - $75,001–$125,000 — 6
   - $125,001–$200,000 — 8
   - Over $200,000 — 10

3. Investable assets:
   - Under $25,000 — 0
   - $25,001–$50,000 — 2
   - $50,001–$100,000 — 4
   - $100,001–$250,000 — 6
   - Over $250,000 — 8

Time and Objective

4. How long will it be until you need to withdraw a significant portion of this money?
    - Less than 3 years   0
    - 3–5 years   4
    - 6–10 years   6
    - 11–15 years   8
    - More than 15 years   10

5. What is your objective with this money?
    - Safety   0
    - Some increase in value   4
    - Beat inflation   8
    - Highest possible returns   10

Risk Tolerance

6. How much volatility are you comfortable with?
    - I never want to see my principle decrease.   0
    - If it decreases for a short time, I'm okay with that.   2
    - I've got a strong stomach; let's go for a ride.   4
    - I'm in it for the long haul. What goes down must go up.   6

7. You bought an investment yesterday for $10,000. Today it's worth $8,000. What do you do?
   - Scream at the top of my lungs: *Get out! Sell!* — 0
   - Move some of the remaining money to something safer. — 2
   - Hold on tight. — 4
   - Buy more. — 6

8. Which are you most comfortable with?
   - +/– 0% — 0
   - +/– 5% — 4
   - +/– 10% — 6
   - +/– 20% — 8
   - +/– more % — 10

Investment Knowledge and Experience

9. Current knowledge of investments:
   - None — 0
   - Basic — 2
   - Moderate — 4
   - Advanced — 6
   - Expert — 8

10. Expected return on this investment:
    - 0–2%   0
    - 2–4%   2
    - 4–6%   4
    - 6–8%   6
    - More than 8%   8

Your score

1.	2.	3.	4.	5.
6.	7.	8.	9.	10.
			Total	

What does your Investor Profile score mean?

Score	Investor Profile	Style
0–10	Conservative	Preservation of capital is key to you. Keep your money in savings or GICs.
11–34	Moderate	Your goals may be quickly approaching, or you are cautious of fluctuation. Keep most of your money safe in portfolios heavily weighted in income-class investments.
35–59	Balanced	Middle of the road; you can stomach a few bumps, and you are hoping to maintain your purchasing power over the long term.
60–74	Growth	You are looking to stay ahead of inflation and taxes. You have a longer time to invest; you can afford some ups and downs, and you are comfortable with that.
75–84	Aggressive	You are comfortable taking a lot of risk. You understand market movements, and you like to buy when others are selling because you have a lot of time on your side. You don't worry about market cycles.

Note: If your answer to Question 4 was 0–3 years, you are strongly advised to choose a conservative approach.

Appendix 2
RECOMMENDED RESOURCES

♣ **Recommended Websites**

Ernst & Young calculators:

- http://www.ey.com/CA/en/Services/Tax/Tax-Calculators-2016-Personal-Tax
- http://www.ey.com/CA/en/Services/Tax/Tax-Calculators-2016-RRSP-Savings

TransUnion

https://www.transunion.ca/ca

Equifax

http://www.consumer.equifax.ca/home/en_ca

CMHC

https://www.cmhc-schl.gc.ca/en/co/buho/

Genworth

http://genworth.ca/en/index.aspx

MFDA Leverage Risk Disclosure

http://www.mfda.ca/regulation/notices/MR-0074.pdf

Mint

https://www.mint.com/

CDIC

http://www.cdic.ca/en/Pages/default.aspx

♦ Recommended Books

The Wealthy Barber by David Chilton

Rich Dad Poor Dad by Robert Kiyosaki

OPM: Other People's Money by Michael Lechter

Think and Grow Rich by Napoleon Hill

The Millionaire Next Door by Thomas Stanley and William Danko

Personal Finance for Dummies by Eric Tyson

INDEX

A

Accessibility, 34, 72–73, 75–76, 78–79
Account nicknames, 10
Advisors, financial, 9, 14, 152
Asset allocation, 84–85
Assets, 16–18, 17, 67
Automatic investment plan, 82, 86

B

Bank statements, 46
Banks, 45–46, 73
Berkshire Hathaway, 80
Bonds, 76–77
Budgets, 21–25, 54
Buffett, Warren, 80

C

Canada Pension Plan (CPP), 84–85, 96, 101
Canadian Deposit Insurance Corporation (CDIC), 74
Canadian Mortgage and Housing Corporation (CMHC), 141–142
Capital, 43
    See also Net worth
Cash, 53–56, 72–73, 75, 76, 79
Cash flow, 14–16, 19, 21–22, 26, 103, 168
Cash flow statements, 14–16, 21–22
Chilton, David, 64
Collateral, 43
Comforts lifestyle, xv, 172–173
Compound growth, 83–84, 85, 86
Consolidation loans, 58, 167, 168
    See also Home-equity plans
Corporate executors, 119–120
Cost of living, xii, 33, 75

Credit
    capacity, 43, 44
    credit cards, 52–56, 58
    credit protection, 126–127
    credit rating, 5, 39, 41, 47, 126
    credit reports, 5, 39–41, 47
    credit vehicles, 51–58
    Five Cs of, 42–44
    home-equity plans, 29, 56–58
    loans, 51–52, 56–58
    mortgages, 56–58, 143–146
    TDSR and GDSR, 44–45
Critical illness coverage, 127–128

D

Debt, 65–66
Dependent lifestyle, xiii–xiv, 166–167
Disability coverage, 4–5, 127–128
Diversification, 84–85
Dollar-cost averaging, 82, 102

E

Emergency cash reserve, 10, 61–67
Equifax, 39–45, 47
Equities, 77
Equity plans, 29, 56–58
    See also Consolidation loans

Ernst and Young tax calculators, 89–90, 131, 181
Estate planning
    companies, 119–120
    estate tax, 121–122
    executors, 119–121
    funeral planning, 121
    power of attorney, 118
    wills, 117–118
Executors, 119–121
Expenses, 14, 19–20, 26

F

Financial advisors, 9, 14, 152
Financial freedom number (FFN), 99–102
Financial planning, 4–10, 149–153
Financial statements. *See* Personal financial statements
Five Cs of credit, 42–44
Fixed-income mutual funds, 76–77
Funeral planning, 121

G

Genworth Financial, 141
Goal triangle, 30–32, 140–141
Goals, 9–10, 21, 30–32, 140–141, 151

Gross debt service ratio (GDSR), 44–45
Guaranteed investment certificates (GICs), 34, 72–76, 79, 132

H

Home Buyer's Plan (HBP), 93–95
Home ownership
    benefits of, 143
    down payment for, 140–141, 142
    Home Buyer's Plan (HBP), 93–95
    mortgage strategies, 143–146
    renting *vs.*, 139–140
    title insurance and, 141–142
Home-equity plans, 29, 56–58
    See also Consolidation loans

I

Identity theft, 42
Income, 14, 16, 26
Income taxes. *See* Taxes
Independent lifestyle, xiv, 168–169
Inflation, 33, 75
    See also Cost of living
Instalment credit vehicles, 51–52

Insurance
    credit protection, 126–127
    critical illness coverage, 127–128
    disability coverage, 4–5, 127–128
    group policies, 126
    individual policies, 125–126
    underwriting, 126
Interest rates, 31, 72, 75
Investing
    accessibility and, 72–73, 75–76, 78–79
    compound growth and, 83–84, 85, 86
    diversification and, 84–85
    dollar-cost averaging and, 82
    guaranteed investment certificates (GICs), 34, 72–76, 79
    leverage and, 133–134, 153
    mutual funds, 76–78, 79, 82
    time and, 83–84
    trade-offs and, 78–79
    umbrellas for, 71–72
    understanding, 79–81
Investor profile questionnaire, 81, 175–179

L

Leverage, 133–134, 153

Liabilities, 16–18, 17

Lifelong Learning Plan (LLP), 93–94

Lifestyle stages, xiii–xvi, 166–174

Liquidity. *See* Accessibility

Loans, 51–52, 56–58

Luxuries lifestyle, xv, 173–174

M

Mortgages, 56–58, 143–146

Mutual Fund Dealers Association of Canada (MFDA), 153

Mutual funds, 76–78, 79, 82

N

Net worth, 14, 16–19, 21–22, 26, 66

Net worth statements, 16–18

Non-registered investments, 72, 86, 132–134

O

Old Age Security (OAS), 96–97

Overdraft protection, 27–29

P

Passive income, 99

Personal financial statements
    about, 13–14
    budgets, 21–25
    cash flow statements, 14–16, 21–22
    net worth statements, 16–18

Planning, financial, 4–10, 149–153

Power of attorney, 118

Probate, 121–122

Protection strategy. *See* Insurance

Q

Quality lifestyle, xv, 170–171

Questionnaires
    investor profile, 81, 175–179
    where's my wedge?, 156–174

R

Registered Education Savings Plan (RESP)
    about, 111–114
    LLP *vs.*, 93
    TFSA and, 113

Registered Retirement Savings
    Plan (RRSP)
    about, 89–92
    automatic savings and, 64, 86
    HBP and, 94–95
    as investment umbrella,
        71–72, 86
    leverage *vs.*, 133–134
    meltdown strategy, 135
    mortgage strategies and,
        144–146
    overcontributions to, 90
    spousal RRSP, 95–98, 136
    strategies for, 102–104
    TFSA *vs.*, 111
Renting, 139–140
Retirement
    financial freedom number
        (FFN), 99–102
    income sources for, 101–102
    planning for, 98–99
    See also Registered Retirement
        Savings Plan (RRSP)
Retirement Income Fund (RIF),
    92–93
Revolving credit vehicles, 52–56
Risk management, 32–35

S

Savings, 61–67, 110
SMART goals, 9–10
Spousal RRSP, 95–98, 136
Standard of living, xiii–xvii, 32,
    66–67
Stocks, 77

T

Taxes
    about, 131
    calculator websites, 181
    capital gains, 133, 134
    income splitting strategy, 136
    minimizing, 132–134
    Notice of Assessment (NOA)
        and, 90–91
    RRSP *vs.* leverage, 133–134
    RRSPs and, 89–92
Tax-Free Savings Account
    (TFSA)
    about, 107–111
    automatic savings and, 64, 86
    contribution limits and,
        108–109
    investing and, 107
    as investment umbrella, 72, 86

RESP and, 113
RRSP *vs.*, 111
Term deposits, 73
See also Guaranteed investment certificates (GICs)
Title insurance, 141–142
Total debt service ratio (TDSR), 44–45
TransUnion, 39–45, 47

W

The Wealthy Barber (Chilton), 64
Where's my wedge?, 156–174
Wills, 117–118

Made in United States
North Haven, CT
02 May 2022